Julia Quick Syntax Reference

A Pocket Guide for Data Science Programming

Antonello Lobianco

Apress®

Julia Quick Syntax Reference: A Pocket Guide for Data Science Programming

Antonello Lobianco
Nancy, France

ISBN-13 (pbk): 978-1-4842-5189-8   ISBN-13 (electronic): 978-1-4842-5190-4
https://doi.org/10.1007/978-1-4842-5190-4

Managing Director, Apress Media LLC: Welmoed Spahr
Acquisitions Editor: Steve Anglin
Development Editor: Matthew Moodie
Coordinating Editor: Mark Powers

Cover designed by eStudioCalamar

Cover image designed by Freepik (www.freepik.com)

Distributed to the book trade worldwide by Springer Science+Business Media New York, 233 Spring Street, 6th Floor, New York, NY 10013. Phone 1-800-SPRINGER, fax (201) 348-4505, e-mail orders-ny@springer-sbm.com, or visit www.springeronline.com. Apress Media, LLC is a California LLC and the sole member (owner) is Springer Science + Business Media Finance Inc (SSBM Finance Inc). SSBM Finance Inc is a **Delaware** corporation.

For information on translations, please e-mail editorial@apress.com; for reprint, paperback, or audio rights, please email bookpermissions@springernature.com.

Apress titles may be purchased in bulk for academic, corporate, or promotional use. eBook versions and licenses are also available for most titles. For more information, reference our Print and eBook Bulk Sales web page at http://www.apress.com/bulk-sales.

Any source code or other supplementary material referenced by the author in this book is available to readers on GitHub via the book's product page, located at www.apress.com/9781484251898. For more detailed information, please visit http://www.apress.com/source-code.

Printed on acid-free paper

Table of Contents

About the Author

Antonello Lobianco, PhD is a research engineer employed by a French Grande école (Polytechnic University). He works on biophysical and economic modeling of the forest sector and is responsible for the Lab Models portfolio. He uses C++, Perl, PHP, Python, and Julia. He teaches environmental and forest economics at the undergraduate and graduate levels and modeling at the PhD level. He has been following the development of Julia as it fits his modeling needs, and he is the author of several Julia packages (search for sylvaticus on GitHub for more information).

About the Technical Reviewer

Germán González-Morris is a polyglot software architect/engineer with 20+ years in the field. He has knowledge of Java (EE), Spring, Haskell, C, Python, and JavaScript, among others. He works with web distributed applications. Germán loves math puzzles (including reading Knuth) and swimming. He has tech reviewed several books, including an application container book (Weblogic), as well as titles covering various programming languages (Haskell, TypeScript, WebAssembly, Math for Coders, and RegExp, for example). You can find more information on his blog (https://devwebcl.blogspot.com/) or Twitter account (@devwebcl).

Acknowledgments

This work has been supported by the French National Research Agency through the Laboratory of Excellence, ARBRE, part of the "Investissements d'Avenir" program (ANR 11 – LABX-0002-01).

I want to thank Germán González-Morris for his valuable help in finding errors in the code and improving the description of the language. I want to also thank Mark Powers, the Apress coordinating editor, for his numerous "check ins" that pushed me to continue and finish the book.

This has been possible thanks to the understanding and support of my family.

Introduction

This Julia quick syntax reference book covers the main syntax elements of the Julia language as well as some of its more important packages.

The first chapter explains the basic skills needed to set up the software you need to run and develop Julia programs, including managing Julia packages.

Chapter 2 presents the many predefined types (integers, strings, arrays, etc.) and the methods to work with them. Memory and copy issues are also presented in this chapter, together with an important discussion about the implementation of the various concepts of *missingness*.

After the basic data types have been introduced, Chapter 3 deals with how to organize them in a sequence of logical statements to compose your program. Control flow, functions, blocks, and scope are all discussed in this chapter.

In Chapter 4, we extend our discussion to custom types—in Julia, both primitive and composite types can be custom-defined—and to their organization in the program, either using inheritance or composition. This chapter will be of particular use to readers accustomed to other object-oriented programs, in order to see how to apply object-oriented concepts in Julia.

Chapter 5 explains how to retrieve the inputs needed by your program from a given source (the terminal, a text/CSV/Excel/JSON file, or a remote resource) and conversely, to export the outputs of your program.

In Chapter 6, we discuss a peculiar feature of Julia, that is, the possibility to manipulate the code itself after it has been parsed, but before it is compiled and run. This paves the way to powerful macro programming. We discuss it and present the concepts of *symbols* and *expressions* in Chapter 6.

Julia is a relatively new language, and while the package ecosystem is developing extremely rapidly (as most packages can be written directly in the Julia language alone), it is highly likely that you will still need libraries for which a direct port to Julia is not yet available. Conversely, your main workflow may be in another, slower, high-level language and you may want to use Julia to implement some performant-critical tasks. Chapter 7 shows how to use C, C++, Python, and R code and their relative libraries in Julia and, conversely, embed Julia code in Python or R programs.

The following chapter (Chapter 8) gives a few recommendations for writing efficient code, with runtime performances comparable to compiled languages. We also deal here with *programmer's efficiency*, discussing profiling and debugging tools and with a short introduction to runtime exceptions.

This completes the discussion of the *core* of the language. Julia, however, has been designed as a thin language where most features are provided by external packages, either shipped with Julia itself (a sort of Julia Standard Library) or provided by third parties.

Therefore, the second part of the book deals with this Julia package ecosystem. Chapter 9 introduces the main packages for working with numerical data: storage with data structure packages like `DataFrames` and `IndexedTables`; munging with `DataFramesMeta`, `Query`, and `Pipe`; and visualization with the `Plot` package.

If Chapter 9 deals with processing numerical data, Chapter 10 deals with mathematical libraries for more theoretical work. `JuMP` is an acclaimed "algebraic modeling language" for numerical optimization (and can be in itself the primary reason to learn about Julia). We present two complete examples with linear and non-linear models. The second model is then rewritten to be analytically resolved with `SymPy`, which is a library for symbolic computation, e.g. the analytical resolution of derivatives, integrals, and equations (and systems of equations). Chapter 10 ends with a presentation of `LsqFit`, a powerful and versatile library to fit data. Finally, Chapter 11 concludes the book with a series of tools that are of more general use, like composing dynamic documents with Wave, dealing with ZIP files with ZipFile, and exposing a given Julia model on the web with Interact and Mux. Examples given in the text are intentionally trivial. They are minimal working examples of the syntax for the concepts they explain. If you are looking for recipes directly applicable to your domain, a "cookbook" kind of book may be more convenient.

While each package has been tested specifically with Julia 1.2 and 1.3-rc4, thanks to the Julia developers' commitment to a stable API, they should remain relevant for the entire 1.x series. Concerning third-party packages, we report the exact version we tested our code with. The section entitled "Using the Package Manager" in Chapter 1 explains how to update a package to a given version if subsequent versions of the package break the API.

Is such cases, please report the problem to us using the form at https://julia-book.com. We will regularly publish updates and errata on this site, where a discussion forum focused on the book is also available.

PART I

Language Core

CHAPTER 1

Getting Started

1.1 Why Julia?

With so many programming languages available, why create yet another one? Why invest the time to learn Julia? Is it worth it?

One of the main arguments in favor of using Julia is that it contributes to improving a trade-off that has long existed in programming—fast coding versus fast execution.

On the one side, Julia allows the developer to code in a dynamic, high-level language similar to Python, R, or MATLAB, interacting with the code and having powerful expressivity (see Chapter 6, for example).

On the other side, with minimum effort, developers can write programs in Julia that run (almost) as fast as programs written in C or FORTRAN.

Wouldn't it be better, though, to optimize existing languages, with their large number of libraries and established ecosystems, rather than create a new language from scratch?

Well, yes and no. Attempts to improve runtime execution of dynamic languages are numerous. PyPy (`https://pypy.org`), Cython (`https://cython.org`), and Numba (`https://numba.pydata.org`) are three notable examples for the Python programming language. They all clash with one fact: Python (and, in general, all the current dynamic languages) was designed before the recent development of just-in-time (JIT) compilers, and hence it offers features that are not easy to optimize. The optimization tools either fail or require complex workarounds in order to work.

© Antonello Lobianco 2019
A. Lobianco, *Julia Quick Syntax Reference*, https://doi.org/10.1007/978-1-4842-5190-4_1

Conversely, Julia has been designed from the ground up to work with JIT compilers, and the language features—and their internal implementations—have been carefully considered in order to provide the programmer with the expected productivity of a modern language, all while respecting the constraints of the compiler. The result is that Julia-compliant code is guaranteed to work with the underlying JIT compiler, producing in the end highly optimized compiled code.

ℹ The Shadow Costs of Using a New Language

If it is true that the main "costs" of using a new language relate to learning the language and having to abandon useful libraries and comfortable, feature-rich development editors that you are accustomed to, it is also true that in the Julia case these costs are mitigated by several factors:

- The language has been designed to syntactically resemble mainstream languages (you'll see it in this book!). If you already know a programming language, chances are you will be at ease with the Julia syntax.

- Julia allows you to easily interface your code with all the major programming languages (see Chapter 7, "Interfacing Julia with Other Languages"), hence reusing their huge sets of libraries (when these are not already ported to Julia).

- The development environments that are available—
  e.g., Juno (`https://junolab.org`), IJulia Jupiter
  kernel (`https://github.com/JuliaLang/
  IJulia.jl`), and VSCode Julia plugin (`https://
  github.com/JuliaEditorSupport/julia-
  vscode`)—are frankly quite cool, with many common
  features already implemented. They allow you to be
  productive in Julia from the first time you code with it.

Apart from the breakout in runtime performances from traditional
high-level dynamic languages, the fact that Julia was created from scratch
means it uses the best, most modern technologies, without concerns over
maintaining compatibility with existing code or internal architectures.
Some of the features of Julia that you will likely appreciate include built-in
Git-based package manager, full code introspection, multiple dispatches,
in-core high-level methods for parallel computing, and Unicode
characters in variable names (e.g., Greek letters).

Thanks to its computational advantages, Julia has its natural roots
in the domain of scientific, high-performance programming, but it is
becoming more and more mature as a general purpose programming
language. This is why this book does not focus specifically on the
mathematical domain, but instead develops a broad set of simple,
elementary examples that are accessible even to beginner programmers.

1.2 Installing Julia

Julia code can be run by installing the Julia binaries for your system
available in the download section (`http://julialang.org/downloads/`) of
the Julia Project website (`https://julialang.org`).

The binaries ship with a Julia interpreter console (aka, the "REPL"—Read, Eval, Print, Loop), where you can run Julia code in a command-line fashion.

For a better experience, check out an Integrated Development Environment, for example, Juno (`http://junolab.org/`) an IDE based on the Atom (`https://atom.io`) text editor, or IJulia (`https://github.com/JuliaLang/IJulia.jl`), the Julia Jupiter (`http://jupyter.org/`) backend.

Detailed setup instructions can be found on their respective sites, but in a nutshell, the steps are pretty straightforward.

- For Juno:

    - Install the main Julia binaries first.

    - Download, install, and open the Atom text editor (`https://atom.io`).

    - From within Atom, go to the Settings ➤ Install panel.

    - Type `uber-juno` into the search box and press Enter. Click the Install button on the package with the same name.

- For IJulia:

    - Install the main Julia binaries first.

    - Install the Python-based Jupyter Notebook server using the favorite tools of your OS (e.g., the Package Manager in Linux, the Python spip package manager, or the Anaconda distribution).

    - From a Julia console, type `using Pkg; Pkg.update();Pkg.add("IJulia");Pkg.build("IJulia")`.

- The IJulia kernel is now installed. Just start the notebook server and access it using a browser.

You can also choose, at least to start with, not to install Julia at all, and try instead one of the online computing environments that support Julia. For example, JuliaBox (`https://juliabox.com/`), CoCalc (`https://cocalc.com/doc/software-julia.html`), Nextjournal (`https://nextjournal.com`), and Binder (`https://mybinder.org`).

💡 Some tricks for Juno and IJulia

- Juno can:

  - Enable block selection mode with `ALT` + `SHIFT`.

  - Run a selection of code by selecting it and either selecting Run Block or typing `SHIFT` + `Enter` on Windows and Linux or `CMD` + `Enter` on Mac.

  - Comment/uncomment a block of code with `CTRL` + `/` (Windows and Linux) or `CMD` + `/` (Mac).

- IJulia:

  - Check out the many keyboard shortcuts available from Help ➤ Keyboard Shortcuts.

  - Need to run Julia in a computational environment for a team or a class? Use JupyterHub (`https://github.com/jupyterhub/jupyterhub`), the multi-user solution based on Jupyter.

1.3 Running Julia

There are many ways to run Julia code, depending on your needs:

1. Julia can run interactively in a console. Start `julia` to obtain the REPL console, and then type the commands there (type `exit()` or use CTRL+D when you are finished).

2. Create a script, i.e. a text file ending in `.jl`, and let Julia parse and run it with `julia myscript.jl [arg1, arg2,..]`.

   Script files can also be run from within the Julia console. Just type `include("myscript.jl")`.

3. In Linux or on MacOS, you can instead add at the top of the script the location of the Julia interpreter on your system, preceded by `#!` and followed by an empty row, e.g. `#!/usr/bin/julia` (You can find the full path of the Julia interpreter by typing `which julia` in a console.). Be sure that the file is executable (e.g., `chmod +x myscript.jl`).

   You can then run the script with `./myscript.jl`.

4. Use an Integrated Development Environment (such as those mentioned), open a Julia script, and use the run command specific to the editor.

You can define a global (for all users of the computer) and local (for a single user) Julia file that will be executed at any startup, where you can for example define functions or variables that should always be available. The location of these two files is as follows:

- Global Julia startup file: [JULIA_INSTALL_FOLDER]\
  etc\julia\startup.jl (where JULIA_INSTALL_FOLDER
  is where Julia is installed)

- Local Julia startup file: [USER_HOME_FOLDER]\.julia\
  config\startup.jl (where USER_HOME_FOLDER is
  the home folder of the local user, e.g. %HOMEPATH% in
  Windows and ~ in Linux)

Remember to use the path with forward slashes (/) with Linux. Note that the local config folder may not exist. In that case, just create the config folder as a .julia subfolder and start the new startup.jl file there.

💡 Julia keeps all the objects created within the same work session in memory. You may sometimes want to free memory or "clean up" your session by deleting no longer needed objects. If you want to do this, just restart the Julia session (you may want to use the trick mentioned at the end of Chapter 3) or use the Revise.jl (https://github.com/timholy/Revise.jl) package for finer control.

You can determine which version of Julia you are using with the versioninfo() option (within a Julia session).

1.4 Miscellaneous Syntax Elements

Julia supports single-line (#) and multi-line (#= [...] =#) comments. Multi-line comments can be nested and appear anywhere in the line:

```julia
println("Some code..")                                    JULIA
#=
  Multiline comment
  #= nested multiline comment =#
  Still a comment
=#
println(#= A comment in the middle of the line =# "This is a
code") # Normal single-line comment
```

You don't need to use semicolons to indicate the end of a statement. If they're used, semicolons will suppress the command output (this is done automatically in scripting mode). If the semicolon is used alone in the REPL, it allows you to switch to the OS command shell prompt in order to launch a system-wide command.

Blocks don't need to be surrounded by parentheses, but they do require the keyword end at the close.

🛇 While indentation doesn't carry any functional meaning in the language, empty spaces sometimes are important. For example, function calls must uses parentheses with the inputs strictly attached to the function name, e.g.:

```text
println (x)   # rise an ERROR                              TEXT

println(x)    # OK
```

In Julia, variable names can include a subset of Unicode symbols, allowing a variable to be represented, for example, by a Greek letter.

In most Julia development environments (including the console), to type a Greek letter, you use a LaTeX-like syntax. This involves typing \, then the LaTeX name for the symbol (e.g. \alpha for α), and then pressing Tab to confirm. Using LaTeX syntax, you can also add subscripts, superscripts, and decorators.

All the following are valid, if not crazy, variable names: x_1, x̃, α, y1, $y^{(a+b)}$, y2, ▮▮▮, and ◔.

Note, however, that while you can use y2 as a variable name, you can't use 2y, as the latter is automatically interpreted as 2 ∗ y. Together with Unicode, this greatly simplifies the transposition in computer code of mathematical equations.

If you come from a language that follows a zero-indexing standard (such as C or Python), one important point to remember is that Julia arrays are one-based indexed (counting starts from 1 and not 0). There are ways to override this behavior, but in many cases doing so probably would do more harm than good.

1.5 Packages

Julia developers have chosen an approach where the core of Julia is relatively light, and additional functionality is usually provided by external "packages".

Julia binaries ship with a set of these packages (think to it as a "Standard Library") and a powerful package manager that is able to download (typically directly from GitHub repositories), pre-compile, update, and solve dependencies, all with a few simple commands.

While *registered* packages can be installed simply by using their name, unregistered packages need their source location to be specified. At the time of this writing, over 2,400 registered packages have been published.

Knowing how packages work is essential to efficiently working in Julia, and this is why I have chosen to introduce package management early in the book and complement the book with a discussion of some common packages.

1.5.1 Using the Package Manager

There are two ways to access package management features, interactively and as an API from within other Julia code:

- The interactive way is to type] in the REPL console to enter a "special" pkg mode. The prompt will then change from julia> to (vX.Y) pkg>, where vX.Y is the current Julia version.

  You can then run any package manager commands or go back to the normal interpreter mode with BACKSPACE.

- The API way is to import the pkg module into your code (using Pkg) and then run Pkg.command(ARGS). Obviously, nothing inhibits you from using the API approach in an interactive session, but the special package mode has tab completion and other goodies that make it more comfortable to use.

- Note that the two interfaces are not 100% consistent, with the API interface being slightly more stringent.

Some of the useful package commands are explained in the following list:

- `status`: Retrieves a list (name and version) of the locally installed packages.

- `update`: Updates the local index of packages and all the local packages to the latest version.

- `add pkgName`: Automatically downloads and installs a given package. For multiple packages use `add Pkg1 Pkg2` or `Pkg.add(["Pkg1","Pkg2"])`.

- `add pkgName#master`, `add pkgName#branchName`, or `add pkgName#vX.Y.Z`: Checks out the master branch of a given package, a specific branch, or a specific release, respectively.

- `free pkgName`: Returns the package to the latest release.

- `rm pkgName`: Removes a package and all its dependent packages that have been installed automatically only for it.

- `add git@github.com:userName/pkgName.jl.git`: Checks out a non-registered package from a Git repository (here, it's GitHub).

1.5.2 Using Packages

To access the functionalities of a package, you need to either *use* or *import* it. The difference is as follows:

- Using a package allows you to access the package functions directly. Just include a `using mypackage` statement in the console or at the beginning of the script.

- Importing a package does the same, but helps in keeping the namespace clean, as you need then to refer to the package functions using their full names, as myPkg.myFunction. You can use aliases or choose to import only a subset of functions (that you can then access directly).

For example, to access the function plot(), which is available in the package Plots, you can do the following (see the "Plotting" section in Chapter 9 for specific plotting issues):

- Access the package function directly with using myPackage:

```julia
using Plots
plot(rand(4,4))
```

- Access the package functions using their full names with import myPackage:

```julia
import Plots
const pl = Plots # This (optionally) creates an an
alias, equivalent to Python `import Plots as pl`.
Declaring it constant may improves the performances.
pl.plot(rand(4,4)) # `Equivalent to Plots.
plot(rand(4,4))`
```

- Access the package functions directly with import myPackage:myfunction:

```julia
import Plots: plot # We can import multiple
functions at once using commas
plot(rand(4,4))
```

Finally, you can also include any Julia source files using this line:

```
include("MyFile.jl") :
```

When that line runs, the included file is completely ran (not only parsed) and any symbol defined there becomes available in the *scope* (the region of code within which a variable is visible) relative to where the include was called.

You can read more about packages in the relevant section (https://julialang.github.io/Pkg.jl/v1/) of the Julia documentation, or by typing help or help COMMAND in pkg mode to get more details on the package manager commands.

! Across this book, I refer to several packages, either in the standard library or third-party packages. When I state that a given function belongs to a given package, remember to add using *PackageName* in order to run the code in the examples (I will not repeat this each time).

1.6 Help System

Julia comes with an integrated help system that retrieves usage information for most functions directly from the source code. This is true also for most third-party packages.

Typing ? in the console leads to the Julia help system, and the prompt changes to help?>. From there, you can search for the function's API.

💡 In non-interactive environment like IJulia notebooks, you can use `?search_term` to access the documentation.

In Juno, you can right-click to open the contextual menu and choose Show Documentation to bring up documentation for the object.

If you don't remember the function name exactly, Julia is kind enough to return a list of similar functions.

While the actual content returned may vary, you can expect to see the following information for each function you query:

- Its signature

- One-line description

- Argument list

- Hints to similar or related functions

- One or more usage examples

- A list of methods (for functions that have multiple implementations)

CHAPTER 2

Data Types and Structures

Julia natively provides a fairly complete and hierarchically organized set of predefined types (especially the numerical ones). These are either *scalar*—like integers, floating-point numbers, and chars,—or *container-like* structures able to hold other objects—like multidimensional arrays, dictionaries, sets, etc.

In this chapter, we discuss them, and in the Chapter 4, where we cover Julia custom types, we consider their hierarchical organization.

Every value (even the primitive ones) has its own unique type. By convention types start with a capital letter, such as Int64 or Bool. Sometimes (such as for all container-like structures and some non-container ones), the name of the type is followed by other parameters inside curly brackets, like the types of the contained elements or the number of dimensions. For example, Array{Int64,2} would be used for a two-dimensional array of integers.

In Julia terminology, these are referred as to *parametric types*. In this book, we will use T as a placeholder to generically indicate a type.

There is no division between object and non-object values. All values in Julia are true objects having a type. Only values, not variables, have types. Variables are simply names bound to values. The :: operator can be used to attach type annotations to expressions and variables in programs. There are two primary reasons to do this:

© Antonello Lobianco 2019
A. Lobianco, *Julia Quick Syntax Reference*, https://doi.org/10.1007/978-1-4842-5190-4_2

- As an assertion to help confirm that your program works the way you expect.

- To provide extra type information to the compiler, which can then, in some cases, improve performance.

2.1 Simple Types (Non-Containers)

Individual characters are represented by the Char type, e.g. a = 'a' (Unicode is fully supported). Boolean values are represented by the Bool type, whose unique instances are true and false.

⊞ In Julia, single and double quotes are not interchangeable. Double quotes produce a char (e.g., x = 'a'), whereas single quotes produce a string (e.g., x = "a").

While the Boolean values true and false in a integer context are automatically cast and evaluated as 1 and 0, respectively, the opposite is not true: if 0 [...] end would indeed rise a non-Boolean (Int64) used in Boolean context TypeError.

The "default" integer type in Julia is Int64 (there are actually 10 different variants of integer types), and it is able to store values between -2^63 and 2^63-1. Similarly, the "default" floating-point type is Float64. Complex numbers (Complex{T}) are supported through the global constant im, representing the principal square root of -1.

A complex number can then be defined as a = 1 + 2im. The mathematical focus of Julia is evident by the fact that there is a native type even for exact ratios of integers, Rational{Int64}, whose instances can be constructed using the // operator:

```
a = 2 // 3
```

2.1.1 Basic Mathematic Operations

All standard basic mathematical arithmetic operators are supported in the obvious way (+ , - , * , /). To raise a number to a power, use theˆ operator (e.g., a = 3^2). Natural exponential expressions (i.e., with the Euler's number e as a base) are created with a = exp(b) or by using the global constant e (This is not the normal letter e, but the special Unicode symbol for the Euler's number: type `\euler` + `TAB` in the REPL to obtain it or use MathConstants.e.). Integer divisions are implemented with the ÷ operator (`\div`) and their remainders are given using the "modulo" % operator, as follows:

```
a = 3 % 2
```

The pi constant is available as a global constant pi or π (`\pi`).

2.1.2 Strings

The String type in Julia can be seen in some ways as a specialized array of individual chars (for example, strings support indexing or looping over their individual letters like an Array would). Unlike arrays, strings are immutable (a="abc"; a[2]='B' would raise an error).

A string on a single row can be created using a single pair of double quotes, while a string on multiple rows can use a triple pair of double quotes:

```
a = "a string"                                          JULIA
b = "a string\non multiple rows\n"
c = """
a string
on multiple rows
"""

a[3] # Returns 's'
```

Julia supports most typical string operations. For example:

- `split(s, " ")` defaults to whitespace

- `join([s1,s2], "")`

- `replace(s, "toSearch" => "toReplace")`

- `strip(s)` Removes leading and trailing whitespace

To convert strings representing numbers to integers or floats, use `myInt = parse(Int,"2017")`. To convert integers or floats to strings, use `myString = string(123)`.

Concatenation

There are several ways to concatenate strings:

- Using the concatenation operator: `*`

- Using the `string` function: `string(str1,str2,str3)`

- Using interpolation, that is combining string variables using the dollar sign: `a = "$str1 is a string and $(myObject.int1) is an integer"` (note the use of parentheses for larger expressions)

 While both the `string` function and the use of interpolation automatically cast compatible types (e.g., `Float64` and `Int64`) into strings, the concatenation operator `*` doesn't. Be careful not to mistake the `string` function (with a lowercase s) with the `String` type and the homonymous constructor (with a capital S).

2.2 Arrays (Lists)

Arrays `Array{T,nDims}` are N-dimensional mutable containers. In this section, we deal with one-dimensional arrays (e.g., `Array{Float64,1}`). In the next section, we consider two or more dimensional arrays (e.g., `Array{Float64,2}`).

💡 In Julia, you may also encounter vectors and matrices.

`Vector{T}` is just an alias for a one-dimensional `Array{T,1}` and `Matrix{T}` is just an alias for a two-dimensional `Array{T,2}`.

There are several ways to create an `Array` of type T:

- Column vector (one-dimensions array): a = `[1;2;3]` or a=`[1,2,3]`

- Row vector (in Julia, this is a two-dimensional array where the first dimension is made of a single row; see the next section "Multidimensional and Nested Arrays"): a = `[1 2 3]`

- Empty (zero-elements) arrays:

    - a = `[]` (resulting in a `Array{Any,1}`)

    - a = `T[]`, e.g. a = `Int64[]`

    - Using the constructor explicitly: a = `Array{T,1}()`

    - Using the `Vector` alias: c = `Vector{T}()`

- *n*-element zeros array: a=`zeros(n)` or a=`zeros(Int64,n)`

- *n*-element ones array: a=`ones(n)` or a=`ones(Int64,n)`

- *n*-element array whose content is garbage:
  a=Array{T,1}(undef,n)

- *n*-element array of identical *j* elements: a=fill(j, n)

- *n*-element array of random numbers: a = rand(n)

Arrays are allowed to store heterogeneous types, but in such cases, the array will be of type Any and it will be in general much slower: x = [10, "foo", false].

However, if you need to store a limited set of types in the array, you can use the Union keyword and still have an efficient implementation, e.g., a = Union{Int64,String,Bool}[10, "Foo", false].

a = Int64[] is just shorthand for a = Array{Int64,1}() (e.g., a = Any[1,1.5,2.5] is equivalent to a = Array{Any,1}([1,1.5,2.5])).

❗ a = Array{Int64,1} (without the parentheses) doesn't create an array at all, but just assigns the "data type" Array{Int64,1} to a.

Square brackets are used to access the elements of an array (e.g. a[1]). The slice syntax from:step:to is generally supported and in several contexts will return a (fast) iterator rather than an array. To then transform the iterator into an array, use collect(myIterator), such as:

a = collect(5:10)

Bounding limits are included in the sequence. For example, collect(4:2:8) will return the array [4,6,8] and collect(8:-2:4) will return [8,6,4].

Together with the keyword end, you can use the slice operator to reverse an array, although you are probably better off with the dedicated function reverse(a).

collect(a[end:-1:1])

You can initialize an array with a mix of values and ranges with either y = [2015; 2025:2030; 2100] (note the semicolons!) or the vcat command (which stands for vertical concatenation):

y = vcat(2015, 2025:2030, 2100)

The following functions are useful when working with arrays:

💡 By convention, functions that end with an exclamation mark will modify the first of their arguments.

- push!(a,b): Push an element to the end of a. If b is an array, it will be added as a single element to a (equivalent to Python append).

- append!(a,b): Append all the elements of b to a. If b is a scalar, obviously push! and append! are interchangeable. Note that a string is treated as an array by append!. It is equivalent to Python extend or +=.

- c = vcat(1,[2,3],[4,5]): Concatenate arrays.

- pop!(a): Remove an element from the end.

- popfirst!(a): Remove an element from the beginning (left) of the array.

- deleteat!(a, pos): Remove an element from an arbitrary position.

- pushfirst!(a,b): Add b to the beginning (left) of a (and no, sorry, appendfirst! doesn't exist).

- `sort!(a)` or `sort(a)`: Sort a (depending on whether you want to modify the original array).

- `unique!(a)` or `unique(a)`: Remove duplicates (depending on whether you want to modify the original array).

- `reverse(a)` or `a[end:-1:1]`: Reverse the order of elements in the array.

- `in(b, a)`: Check for the existence of b in a. Also available as an operator: `if b in a [...] end`.

- `length(a)`: Get the length of a.

- `a...` (the "splat" operator): Convert the values of an array into function parameters (used inside function calls; see the next line).

- `maximum(a)` or `max(a...)`: Get the maximum value (max returns the maximum value between the given arguments).

- `minimum(a)` or `min(a...)`: Get the minimum value (min returns the maximum value between the given arguments).

- `sum(a)`: Return the summation of the elements of a.

- `cumsum(a)`: Return the cumulative sum of each element of a (i.e., it returns an array).

- `empty!(a)`: Empty an array (works only for column vectors, not for row vectors).

- `b = vec(a)`: Transform row vectors into column vectors.

- `shuffle(a)` or `shuffle!(a)`: Random-shuffle the elements of a (requires using `Random` before).

- `isempty(a)`: Check if an array is empty.

- `findall(x -> x == value, myArray)`: Find a value in an array and return its indexes. This is a bit tricky. The first argument is an anonymous function (see Chapter 3) that returns a Boolean value for each value of `myArray`, and then `findall()` returns the index positions.

- `deleteat!(myarray, findall(x -> x == myUnwantedItem, myArray))`: Delete a given item from a list.

- `enumerate(a)`: Get (`index,element`) pairs, i.e. return an iterator to tuples, where the first element is the index of each element of the array `a` and the second is the element itself.

- `zip(a,b)`: Get (`a_element, b_element`) pairs, i.e., return an iterator to tuples made of elements from each of the arguments (e.g., `zip(names,sex,age)` will result in something like `[("Marc",'M',18),("Anne", 'F',16)]`). Note that, like enumerate, it is an iterator that is returned.

2.2.1 Multidimensional and Nested Arrays

In this section, we deal with multi-dimensional arrays (e.g., a two-dimensional array `Array{T,2}` or `Matrix{T}`) and nested arrays, as an array of arrays `Array{Array{T,1},1}` (i.e., the main structure remains a one-dimensional array, but the individual elements are themselves arrays). The main difference between a *matrix* and an *array of array* is that, with a matrix, the number of elements on each column (row) must be the same and rules of linear algebra apply. As in matrix algebra, the first dimension is interpreted as the vertical dimension (rows) and the second dimension is the horizontal one (columns).

Multi-dimensional arrays `Array{T,N}` can be created similarly to as one-dimensional ones (in fact, the later are just specific cases of the former):

- By column: a = `[[1,2,3] [4,5,6]]`. Elements of the first column, elements of the second column, and so on. Note that this is valid only if you write the matrix in a single line.

- By row: a = `[1 4; 2 5; 3 6]`. Elements of the first row, elements of the second row.

- Empty (zero-element) arrays:

  - Using the constructor explicitly: a = `Array{T}(undef, 0, 0, 0)`

  - Using the `Matrix` alias: a = `Matrix{T}()`

- n,m,g-element zeros array: a = `zeros(n,m,g)` or a = `zeros(Int64,n,m,g)`

- n,m,g-element ones array: a = `ones(n,m,g)` or a = `ones(Int64,n,m,g)`

- n,m,g-element array whose content is garbage: a = `Array{T,3}(undef,n,m,g)`

- n,m,g-element array of identical j elements: a = `fill(j,n,m,n)`

- n,m,g-element array of of random numbers: a = `rand(n,m,g)`

Multidimensional arrays often arise from using list comprehension. For example, a = `[3x + 2y + z for x in 1:2, y in 2:3, z in 1:2]` (see Chapter 3 for details about list comprehension).

🛑 *Matrix versus Nested Arrays*

Note this important difference:

- a = [[1,2,3],[4,5,6]] creates a one-dimensional array with two elements (each of those is again a vector).

- a = [[1,2,3] [4,5,6]] creates a two-dimensional array (a matrix with two columns) with six scalar values.

Nested arrays can be accessed with double square brackets, such as a[2][3].

Elements of n-dimensional arrays can be accessed instead with the a[idxDim1,idxDim2,..,idxDimN] syntax (e.g., a[row,col] for a matrix), where again the slice syntax can be used. For example, given that a is a 3x3 matrix, a[1:2,:] would return a 2x3 matrix with all the column elements of the first and second rows.

Note that in order to push elements to an array, they have to be compatible with the array type. The following code would not work:

```
a = [1,2,3]; push!(a,[1 2])
```

Because the first command automatically creates an Array{Int64,1}, i.e., an array of scalars, you then try to push on it an inner array. In order for this work, a must be defined either as a = Any[1,2,3] or more explicitly as a = Union{Int64,Array{Int64,1}}[1,2,3].

Boolean selection is implemented using a Boolean array (possibly multidimensional) for the selection:

```
a = [[1,2,3] [4,5,6]]                                        JULIA
mask = [[true,true,false] [false,true,false]]
```

a[mask] returns a one-dimensional array with 1, 2, and 5. Note that the Boolean selection always results in a flatted array, even if you delete a whole row or a whole column of the original data. It is up to the programmer to reshape the data accordingly.

ⓘ For row vectors, both a[2] and a[1,2] return the second element.

Several functions are particularly useful when working with n-dimensional arrays (on top of those you saw in the previous section):

- size(a): Return a tuple (i.e., an immutable list) with the sizes of the *n* dimensions.

- ndims(a): Return the number of dimensions of the array (e.g., 2 for a matrix).

- reshape(a, nElementsDim1, nElementsDim2,..., nElemensDimN): Reshape the elements of a in a new n-dimensional array with the dimensions given. Returns a DimensionMismatch error if the new dimensions are not consistent with the number of elements held by a.

- dropdims(a, dims=(dimToDrop1,dimToDrop2)): Remove the specified dimensions, provided that the specified dimension has only a single element, e.g., a = rand(2,1,3); dropdims(a,dims=(2)).

- transpose(a) or a': Transpose a one- or two-dimensional array.

⚠ Do not confuse `reshape(a)` with `transpose(a)`:

```julia
julia> a = [[1 2 3]; [4 5 6]]                                          JULIA
2×3 Array{Int64,2}:
 1  2  3
 4  5  6

julia> reshape(a,3,2)
3×2 Array{Int64,2}:
 1  5
 4  3
 2  6

julia> transpose(a)
3×2 LinearAlgebra.Adjoint{Int64,Array{Int64,2}}:
 1  4
 2  5
 3  6
```

- `collect(Iterators.flatten(a))`, vec(a), or a[:]:
  Flatten a multi-dimensional array into a column vector.

ℹ Note that `reshape()`, `transpose()`, and `vec()` perform a *shadow copy*, returning a different "view" of the underlying data (so modifying the original matrix modifies the reshaped, transposed, or flatted matrix). You can use `collect()`to force a *deep copy*, where the new matrix holds independent data (see the section "Memory and Copy Issues" for details).

- `hcat(col1, col2)`: Concatenate horizontally anything that makes sense to concatenate horizontally (vectors, matrices, DataFrames,…).

- `vcat(row1, row2)`: Concatenate vertically anything that makes sense to concatenate vertically (vectors, matrices, DataFrames,…).

2.3 Tuples

Use tuples (`Tuple{T1,T2,...}`) to create a immutable list of elements:

`t = (1,2.5,"a")`

Or without parentheses:

`t = 1,2.5,"a"`

Immutable refers to the fact that once they are created, elements of this data structure cannot be added, removed, or changed (rebound to other objects).

💡 Note that if the element in a tuple is itself mutable (e.g., an `Array`), this is still allowed to mutate (i.e. "internally" change). What is "immutable" is the memory addresses or the actual bits of the various objects in the tuple.

Tuples can be easily unpacked to multiple variables: `var1, var2 = (x,y)`. This is useful, for example, for collecting the values of functions returning multiple values, where the returned object would be a tuple.

ⓘ In contrast to arrays, tuples remain efficient even when hosting heterogeneous types, as the information on the types of each element hosted is retained in the type signature. For example, `typeof(((1,2.5,"a"))` is *not* `Tuple{Any}` (as it would be for an array) but `Tuple{Int64,Float64,String}` (You will see why this matters in Chapter 8.).

Useful tricks:

- Convert a tuple into an array: `t=(1,2,3); a = [t...]` or `a = [i[1] for i in t]` or `a=collect(t)`

- Convert an array into a tuple: `t = (a...,)` (note the comma after the splat operator)

2.4 Named Tuples

Named tuples (`NamedTuple`) are collections of items whose position in the collection (index) can be identified not only by their position but also by their name.

- `nt = (a=1, b=2.5)`: Define a `NamedTuple`

- `nt.a`: Access the elements with the dot notation

- `keys(nt)`: Return a tuple of the keys

- `values(nt)`: Return a tuple of the values

- `collect(nt)`: Return an array of the values

- `pairs(nt)`: Return an iterable of the pairs (key,value). Useful for looping: `for (k,v) in pairs(nt) [...] end`

As with "normal" tuples, named tuples can hold any value, but cannot be modified (i.e., are immutable). They too remain efficient even when hosting heterogeneous types.

ℹ️ Before Julia 1.0, named tuples were implemented in a separate package (`NamedTuple.jl`, at `https://github.com/JuliaData/NamedTuples.jl`.) Now that they are implemented in the Julia core, use of that package implementation is discouraged, unless you have to use packages that still depend on it.

2.5 Dictionaries

Dictionaries (`Dict{Tkey,Tvalue}`) store mappings from keys to values and they have an apparently random sorting.

You can create an empty (zero-element) dictionary with `mydict = Dict()` or specify the key and values types with `Dict{String,Int64}()`. To initialize a dictionary with some values, use:

```
mydict = Dict('a'=>1, 'b'=>2, 'c'=>3)
```

Here are some useful methods for working with dictionaries:

- `mydict[akey] = avalue`: Add pairs to the dictionary.

- `delete!(amydict,'akey')`: Delete the pair with the specified key from the dictionary.

- `map((i,j) -> mydict[i]=j, ['a','b','c'], [1,2,3])`: Add pairs using maps (i.e., from vectors of keys and values to the dictionary).

- `mydict['a']`: Retrieve a value using the key (it raises an error if the key doesn't exist).

- `get(mydict,'a',0)`: Retrieve a value with a default value for a nonexistent key.

- `keys(mydict)`: Return all the keys (the result is an iterator, not an array). Use `collect()` to transform it.

- `values(mydict)`: Return all the values (result is again an iterator).

- `haskey(mydict, 'a')`: Check if a key exists.

- `in(('a' => 1), mydict)`: Check if a given key/value pair exists (that is, if the key exists and has that specific value).

You can iterate through the key and the values of a dictionary at the same time:

```julia
for (k,v) in mydict
    println("$k is $v")
end
```

While named tuples and dictionaries can look similar, there are some important differences between them:

- Name tuples are immutable, while dictionaries are mutable.

- Dictionaries are type-unstable if different types of values are stored, while named tuples remain type-stable:

  - `d = Dict(:k1=>"v1", :k2=>2) #` `Dict{Symbol,Any}`

  - `nt = (k1="v1", k2=2) # NamedTuple{(:k1, :k2),Tuple{String,Int64}}`

- The syntax is a bit less verbose and readable with named tuples: `nt.k1` versus `d[:k1]`.

Overall, named tuples are generally more efficient and should be thought more as anonymous `structs` rather than as dictionaries.

2.6 Sets

Use sets (`Set{T}`) to represent collections of unordered, unique values. Sets are mutable collections.

Here are some methods:

- `s = Set()` or `s = Set{T}()`: Create an empty (zero-element) set.

- `s = Set([1,2,2,3,4])`: Initialize a set with values. Note that the set will then have only one instance of 2.

- `push!(s, 5)`: Add elements.

- `delete!(s,1)`: Delete elements.

- `intersect(set1,set2)`, `union(set1,set2)`, `setdiff(set1,set2)`: Set common operations of intersection, union, and difference.

2.7 Memory and Copy Issues

In order to avoid copying large amounts of data, Julia by default copies only the memory addresses of objects, unless the programmer explicitly requests a so-called "deep" copy or the compiler determines that an actual copy is more efficient.

Use `copy()` or `deepcopy()` when you don't want subsequent modifications of the copied object to apply to the original object.

In detail:

`Equal sign (a=b)`

- This performs a *name binding*, i.e., it binds (assigns) the entity (object) referenced by b to the a identifier (the variable name).

- The possible results are as follows:

    - If b rebinds to some other object, a remains referenced to the original object.

    - If the object referenced by b mutates (i.e., it internally changes), so does (being the same object) those referenced by a.

- If b is immutable and small in memory, under some circumstances, the compiler would instead create a new object and bind it to a, but being immutable for the user this difference would not be noticeable

- As for many high level languages, you don't need to explicitly worry about memory leaks. A garbage collector exists so that objects that are no longer accessible are automatically destroyed.

```
a = copy(b)
```

- This creates a new, "independent" copy of the object and binds it to a. This new object may reference other objects through their memory addresses. In that case, it is their memory addresses that are copied and not the referenced objects.

- The possible results are as follows:

    - If these referenced objects (e.g., the individual elements of the vector) are rebound to some other objects, the new object referenced by a maintains the reference to the original objects.

    - If these referenced objects mutate, so do (being the same objects) the objects referenced by the new object referenced by a.

```
a = deepcopy(b)
```

- Everything is "deep copied" recursively.

The following code snippet highlights the differences between these three methods for "copying" an object:

```julia
julia> a = [[[1,2],3],4] # [[[1, 2], 3], 4]                    JULIA
julia> b = a             # [[[1, 2], 3], 4]
julia> c = copy(a)       # [[[1, 2], 3], 4]
julia> d = deepcopy(a)   # [[[1, 2], 3], 4]
# Rebinds a[2] to an other objects.
# At the same time mutates object a:
julia> a[2] = 40         # 40
julia> b                 # [[[1, 2], 3], 40]
julia> c                 # [[[1, 2], 3], 4]
julia> d                 # [[[1, 2], 3], 4]
# Rebinds a[1][2] and at the same time
# mutates both a and a[1]:
julia> a[1][2] = 30      # 30
julia> b                 # [[[1, 2], 30], 40]
julia> c                 # [[[1, 2], 30], 4]
julia> d                 # [[[1, 2], 3], 4]
# Rebinds a[1][1][2] and at the same time
# mutates a, a[1] and a[1][1]:
julia> a[1][1][2] = 20   # 20
julia> b                 # [[[1, 20], 30], 40]
julia> c                 # [[[1, 20], 30], 4]
julia> d                 # [[[1, 2], 3], 4]
# Rebinds a:
julia> a = 5             # 5
julia> b                 # [[[1, 20], 30], 40]
julia> c                 # [[[1, 20], 30], 4]
julia> d                 # [[[1, 2], 3], 4]
```

You can check if two objects have the same values with == and if two objects are actually the same with === (keep in mind that immutable objects are checked at the bit level and mutable objects are checked for their memory address.):

- Given a = [1, 2]; b = [1, 2]; a == b and a === a are true, but a === b is false.

- Given a = (1, 2); b = (1, 2); all a == b, a === a and a === b are true.

2.8 Various Notes on Data Types

To convert ("cast") an object into a different type, use convertedObj = convert(T,x). When conversion is not possible, e.g., when you're trying to convert a 6.4 Float64 into an Int64 value, an error will be thrown (InexactError in this case).

The const keyword, when applied to a variable (e.g., const x = 5), indicates that the identifier cannot be used to bind objects of a different type. The referenced object can still mutate or the identifier rebound to another object of the same type (but in this later case a warning is issued). Only global variables can be declared const.

You can "broadcast" a function to work over a collection (instead of a scalar) using the dot (.) operator.

For example, to broadcast parse to work over an array, you would use:

```
myNewList = parse.(Float64,["1.1","1.2"])
```

(See more information about broadcasting in Chapter 3.)

2.8.1 Random Numbers

It is easy to obtain pseudo-numbers in Julia:

- Random float in [0,1]: `rand()`

- Random integer in [a,b]: `rand(a:b)`

- Random float in [a,b] with precision to the second digit:
  `rand(a:0.01:b)`

- Random float in [a,b]: `rand(Uniform(a,b))` (requires
  the `Distributions` package)

- Random float in [a,b] using a particular distribution
  (Normal, Poisson,...): `rand(DistributionName`
  `([distribution parameters]))`

You can obtain an array or a matrix of random numbers by simply specifying the requested size to `rand()`. For example, `rand(2,3)` or `rand(Uniform(a,b),2,3)` for a 2x3 matrix.

You can "fix" the seed of the pseudo-random number generator with `import Random:seed!; seed!(1234)`. Doing so, you are still dealing with (pseudo) random numbers, but the sequence that's generated will be the same every time you run the Julia script, i.e., generating reproducible output. If you then ever need to revert to irreproducible randomness, you can use `seed!()` without arguments.

2.8.2 Missing, Nothing, and NaN

Julia supports different concepts of *missingness:*

- nothing (type `Nothing`): This is the value returned by
  code blocks and functions that do not return anything.
  It is a single instance of the singleton type `Nothing`, and
  is closer to C-style `NULL` (sometimes it is referred to as

the "software engineer's null"). Most operations with
nothing values will result in a run-type error. In certain
context, it is printed as #NULL.

- missing (type Missing): Represents a missing value in a
  statistical sense. There should be a value but you don't
  know what it is (so it is sometimes referred to as
  the "data scientist's null"). Most operations with
  missing values will propagate (silently). Containers
  can handle missing values efficiently when they are
  declared of type Union{T,Missing}. The Missing.jl
  (https://github.com/JuliaData/Missings.jl)
  package provides additional methods to handle
  missing elements.

- NaN (type Float64): Represents when an operation
  results in a Not-a-Number value (e.g., 0/0). It is similar
  to missing in that it propagates silently rather than
  resulting in a run-type error. Similarly, Julia also offers
  Inf (e.g., 1/0) and -Inf (e.g., -1/0).

CHAPTER 3

Control Flow and Functions

Now that you've been introduced to the main types and containers available in Julia, it is time to put them to work.

3.1 Code Block Structure and Variable Scope

All the typical flow-control constructs (`for`, `while`, `if/else`, and `do`) are supported. The syntax typically takes this form:

```
<keyword> <condition>                                    JULIA
    ... block content...
end
```

For example:

```
for i in 1:5                                             JULIA
    println(i)
end
```

Note that parentheses around the condition are not necessary, and the block ends with the keyword end.

© Antonello Lobianco 2019
A. Lobianco, *Julia Quick Syntax Reference*, https://doi.org/10.1007/978-1-4842-5190-4_3

Variable Scope

Note that within a block the visibility of a variable changes, and Julia is somehow more restrictive than most other languages in this regard.

When you define a variable (in the REPL, Atom, or IJulia Notebook, for example), you do so in the global scope.

Most blocks (for, while, but notably not if) and functions define new local scopes that inherit from the surrounding scope.

Global variables are inherited only for reading, not for writing. The following snippet will result in an UndefVarrError: code not defined error.

```julia
a =  4
while a > 2
    a -= 1
end
```

To assign to a global variable, you need an explicit global:

```julia
a =  5
while a > 2
    global a
    println("a: $a")
    b = a
    while b > 2
        println("b: $b")
        b -= 1
    end
    a -= 1
end
```

Note that assignment to b is still possible, as b is *not* a global variable, but is a local variable defined in the first `while` block and then inherited in the second one.

With the keyword `local` x you specify that the variable x must be treated as a new local variable, independent of the homonymous global one.

Abusing global variables makes the code more difficult to read, may lead to unintended results, and is often the cause of computational bottleneck. Use them sparingly!

3.2 Repeated Iteration: for and while Loops, List Comprehension, Maps

The `for` and `while` constructs are very flexible. First, the condition can be expressed in many ways: `for i = 1:2`, `for i in anArray`, `while i < 3`...
Second, multiple conditions can be specified in the same loop. Consider this example:

```
for i=1:2,j=2:5                              JULIA
  println("i: $i, j: $j")
end
```

In this case, a higher-level loop starts over the range `1:2`, and then for each element, a nested loop over `2:5` is executed.

`break` and `continue` are supported and work as expected—break immediately aborts the loop sequence, while `continue` immediately passes to the next iteration.

Julia supports other constructs for repeated iteration, namely list comprehension and maps.

List comprehension is essentially a very concise way to write a for loop:

```
[myfunction(i) for i in [1,2,3]]
```

```
[x + 2y for x in [10,20,30], y in [1,2,3]]
```

For example, you could use list comprehension to populate a dictionary from one or more arrays:

```
[mydict[i]=value for (i, value) in enumerate(mylist)]
```

```
[students[name] = sex for (name,sex) in zip(names,sexes)]
```

You can write complex expressions with list comprehension, such as [println("i: $i - j: $j") for i in 1:5, j in 2:5 if i > j], but at this point perhaps it's best to write the loop explicitly.

map applies a function to a list of arguments. The same example of populating a dictionary can also be written (a bit less efficiently) using map:

```
map((n,s) -> students[n] = s, names, sexes)
```

When mapping a function with a single parameter, the parameter can be omitted, as follows:

```
a = map(f, [1,2,3]) is equal to a = map(x->f(x), [1,2,3])
```

3.3 Conditional Statements: if Blocks, Ternary Operator

Conditional statements can be written with the typical if/elseif/else construct:

```
i = 5                                                    JULIA
if i == 1
    println("i is 1")
```

44

```
elseif i == 2
    println("i is 2")
else
    println("is in neither 1 or 2")
end
```

Multiple conditions can be considered using the logical operators: *and* (&&), *or* (||), and *not* (!) (not to be confused with the bitwise operators, & and |).

Note that when the result of the conditional statement can be inferred before the conclusion of the expression evaluation, the remaining parts are not evaluated (they are "short-circuited"). Typical cases are when the first condition of an && operator is false or the conditions of the || operator are true. In such cases, it is useless to evaluate the second condition.

In the same way that list comprehension is a concise way to write for loops, the ternary operator is a concise way to write conditional statements.

The syntax is as follows:

```
a ? b : c
```

This means "if a is true, then execute expression b; otherwise, execute expression c ". Be sure there are spaces around the ? and : operators. As with list comprehension, it's important not to abuse the ternary operator in order to write complex conditional logic.

3.4 Functions

Julia's functions are very flexible. They can be defined inline as f(x,y) = 2x+y, or with their own block introduced using the function keyword:

```
function f(x)                                              JULIA
   x+2
end
```

45

A common third way to define functions it is to create an anonymous function and assign it to a nameplace (we discuss this a bit later).

After a function has been *defined*, you can call it to execute it. Note that, as with most high-level languages, there isn't a separate step for *declaring* the function in Julia.

Functions can even be nested, in the sense that the function definition can be embedded within another function definition, or can be recursive, in the sense that there is a call to itself inside the function definition:

```julia
# A nested function:                                       JULIA
function f1(x)
    function f2(x,y)
        x+y
    end
    f2(x,2)
end
# A recursive function:
function fib(n)
    if n == 0 return 0
    elseif n == 1 return 1
    else
     return fib(n-1) + fib(n-2)
    end
end
```

Within the Julia community, it is considered good programming practice for the call to a function to follow these rules:

- To contain all the elements that the function needs for its logic (i.e., no read access to other variables, except constant globals).

- That the function doesn't change any other part
  of the program that is not within the parameters
  (i.e., it doesn't produce any "side effects" other than
  eventually modifying its arguments).

Following these rules will help you achieve code that is fast, reliable (the output of each function depends uniquely on the set of its inputs), and easy to read and debug.

3.4.1 Arguments

Function arguments are normally specified by position (*positional arguments*). However, if a semicolon (;) is used in the parameter list of the function definition, the arguments listed after that semicolon must be specified by name (*keyword arguments*).

⚠ The function call must respect this distinction, calling positional arguments by position and keyword arguments by name. In other words, it's not possible to call positional arguments by name or keyword arguments by position.

The last arguments (whether positional or keyword) can be specified together with a default value. For example:

- Definition: `myfunction(a,b=1;c=2) = (a+b+c)`
  (definition with two positional arguments and one
  keyword argument)

- Function call: `myfunction(1,c=3)` (calling `(1+1+3)`
  Note that b is not provided in the call, and hence the
  default value is used)

You can optionally restrict the types of argument the function should accept by annotating the parameter with the type:

```
myfunction(a::Int64,b::Int64=1;c::Int64=2) = (a+b+c)
```

The reason to specify parameter types is not so much to obtain speed gains. Julia will try to resolve the possible parameter types and return values, and only in rare cases will it be unable to uniquely determine the return type as a function of the input types. In these cases (type-unstable), specifying the parameter types may help solve type instability.

The most important reason to limit the parameter type is to catch bugs early on, when the function is accidentally called with a parameter type it was not designed to work with. In such cases, if the function was annotated with the allowed parameter type, Julia will return to the user a useful error message instead of silently trying to use that parameter.

However, a common case is when you want the function to process single values (scalars) or vectors of a given parameter. You have two options then:

- You can write the function to treat the scalar and rely then on the dotted notation to broadcast the function at call time (discussed more later).

- Alternatively, you may want to directly deal with this in the function definition. In such cases, you can declare the parameter as being either a scalar type T or a vector T using a union. For example: `function f(par::Union{Float64, Vector{Float64}}) [...]` end. You can then implement the logic you want by checking the parameter type using `typeof`.

Finally, functions in Julia may also accept a variable number of arguments. The splat operator (i.e., the ellipsis ...) can specify a variable number of arguments in the parameter declaration within the function definition and can "splice" a list or an array in the parameters within the function call:

```julia
values = [1,2,3]
function additionalAverage(init, args...)
#The parameter that uses the ellipsis must be the last one
  s = 0
  for arg in args
    s += arg
  end
  return init + s/length(args)
end
a = additionalAverage(10,1,2,3)         # 12.0
a = additionalAverage(10, values ...)   # 12.0
```

3.4.2 Return Value

Providing a return value using the return keyword is optional. By default, functions return the last computed value.

Often, return is used to immediately terminate a function, for example, upon the occurrence of certain conditions. Note that the return value can also be a tuple (returning multiple values at once):

```julia
myfunction(a,b) = a*2,b+2
x,y = myfunction(1,2)
```

3.4.3 Multiple-Dispatch (aka Polymorphism)

When similar logic should be applied to different kinds of objects (i.e., different types), you can write functions that share the same name but have different types or different numbers of parameters (and different implementation). This highly simplifies the Application Programming Interface (API) of your application, as only one name has to be remembered.

When calling such functions, Julia will pick up the correct one depending on the parameters in the call, selecting by default the stricter version.

These different versions are called *methods* in Julia and, if the function is type-safe, dispatch is implemented at compile time and is very fast.

You can list all the methods of a given function by using methods(myfunction).

The multiple-dispatch polymorphism is a generalization of object-oriented runtime polymorphism, where the same function name performs different tasks, depending on which is the object's class. The polymorphism in traditional object-oriented languages is applied only to a single element, whereas in Julia it applies to all the function arguments (it remains true however that object-oriented languages have multiple-parameters polymorphism at compile-time).

I'll go more in depth into multiple-dispatch when dealing with type inheritance in Chapter 4.

3.4.4 Templates (Type Parameterization)

Functions can be specified regarding which types they work with. You do this using *templates:*

```
myfunction(x::T, y::T2, z::T2) where {T <: Number, T2} = x + y + z
```

This function first defines two types, T (a subset of Number) and T2, and then specifies which of these two types each parameter must be.

You can call it with (1,2,3) or (1,2.5,3.5) as a parameter, but not with (1,2,3.5) as the definition of myfunction requires that the second and third parameter must be the same type (whatever that is).

3.4.5 Functions as Objects

Functions themselves are objects and can be assigned to new variables, returned, or nested. Take this example:

```julia
f(x) = 2x # define a function f inline                    JULIA
a = f(2)   # call f and assign the return value to a. `a` is a value
a = f      # bind f to a new variable name. `a` is now a function
a(5)       # call again the (same) function
```

3.4.6 Call by Reference/Call by Value

Julia functions are called using a convention—sometimes known as *call-by-sharing* in other languages—that is somehow in between the traditional *call by reference* (where just a memory pointer to the original variable is passed to the function) and *call by value* (where a copy of the variable is passed, and the function works on this copy).

In Julia, functions work on new local variables, known only inside the function itself. Assigning the variable to another object will not influence the original variable. But if the object bound with the variable is mutable (e.g., an array), the *mutation* of this object will apply to the original variable as well:

```julia
function f(x,y)                                           JULIA
    x = 10
    y[1] = 10
end
x = 1
y = [1,1]
f(x,y) # x will not change, but y will now be [10,1]
```

(See also the "Memory and Copy Issues" section at the end of Chapter 2.)

Functions that change their arguments have their name, by convention, followed by an !. For example, `myfunction!(ref_par, other_pars)`. The first parameter is, still by convention, the one that will be modified.

3.4.7 Anonymous Functions (aka "Lambda" Functions)

Sometimes you don't need to name a function (e.g., when the function is one of the arguments being passed to higher-order functions, like the `map` function). To define anonymous (nameless) functions, you can use the `->` syntax, as follows:

```julia
x -> x^2 + 2x - 1
```

This defines a nameless function that takes an argument, x, and produces `x^2 + 2x - 1`. Multiple arguments can be provided using tuples, as follows:

```
(x,y,z) -> x + y + z
```

You can still assign an anonymous function to a variable this way:

```
f = (x,y) -> x+y
```

⚠️ Do not confuse the single arrow `->`, which is used to define anonymous functions, with the double arrow `=>`, which is used to define a `pair` (e.g., in a dictionary).

3.4.8 Broadcasting Functions

You'll often have a function designed to work with scalars, but you want to apply it repetitively to values within a container, like the elements of an array. Instead of writing for loops, you can rely on a native functionality of Julia, which is to *broadcast* the function over the elements you wish.

Take, for example, the following function:

```julia
f1(a::Int64,b::Int64) = a*b
```

It expects as input two scalars. For example:

```julia
f1(2,3)
```

But what if a and b are vectors (say a=[2,3] and b=[3,4])? You can't directly call the function f1([2,3],[3,4]). The solution is to use the function broadcast(), which takes the original function as its first argument followed by the original function's arguments: broadcast(f1,[2,3],[3,4]).

The output is a vector that holds the result of the original function applied first to (a=2,b=3) and then to (a=3,b=4).

A handy shortcut to broadcast is to use the dot notation, i.e., the original function name followed by a dot: f1.([2,3],[3,4]).

Sometimes the original function natively takes some parameters as a vector, and you want to limit the broadcast to the scalar parameters.

In such cases, you can use the Ref() function to protect the parameters that you don't want to be broadcast:

```julia
f2(a::Int64,b::Int64,c::Array{Int64,1},d::Array{Int64,1}) =
a*b+sum(c)-sum(d)                                              JULIA
f2(1,2,[1,2,3],[0,0,1]) # normal call without broadcast
f2.([1,1,1],[2,2,2],Ref([1,2,3]),Ref([0,0,1])) # broadcast over
the first two arguments only
```

3.5 Do Blocks

We finish the chapter by analyzing do blocks. Do blocks allow developers to define "anonymous" functions that are passed as arguments to outer functions.

For example, you write `f1(f2,x,y) = f2(x+1,x+2)+y`. To use this `f1` function, you first need another function to act as `f2` (the inner function). You could define it as `f2(g,z) = g*z` and then call `f1` as `f1(f2,2,8)`. Every time the first argument is a function, this can be written *a posteriori* with a do block. You can hence obtain the same result using the do block:

```julia
f1(2,8) do i,j
    i*j
end
```

JULIA

This defines `i` and `j` as local variables that are made available to the do block. Their values are determined in the `f1` function (in this case, `i=2+1` and `j=2+2`). The result of the block computation is then made available as the output of the function acting as the first parameter of `f1`. Again, what you do with this value is specified by the definition of the `f1` function (in this case, the value 8 is added to it to make 20, the returned value).

Another typical use of do blocks is within input/output operations, as described in Chapter 5.

3.6 Exiting Julia

To exit a running Julia session, press `CTRL` + `D` (or use `CTRL` + `C` to throw an `InterruptException` when possible, regaining control of the Julia prompt without closing the process).

To exit the script programmatically use `exit(exit_code)` where the default exit code of zeros indicates, by convention, a normal program termination.

Finally, sometimes you'll want to be able to define something your code should run whenever the Julia process exits.

This is exactly the role of the atexit(f) function. It allows you to define a zero-argument function f() to be called upon exit.

You can use it, for example, to start a new "clean Julia session" when a given function is invoked. As an example, type the following function in your global or local startup.jl file:

```julia
function workspace()
    atexit() do
        run(`$(Base.julia_cmd())`)
    end
    exit()
end
```

Now, whenever the workspace() function is invoked, the current Julia process will be closed and a new one will be started (Note that workspace was the name of a function in previous Julia versions that cleaned up the Workspace; it's no longer available. This script replicates its functionality, albeit at the cost of forcing a restart of the whole Julia process.).

CHAPTER 4

Custom Types

Chapter 2, "Data Types and Structures," discussed built-in types, including containers. In this chapter, we move to study how to create user-defined types.

ⓘ *Type versus Structure*

Let's clarify these two terms in the Julia language context.

A *type* of an object, in plain English, refers to the set of characteristics that describe the object. For example, an object of type *sheet* can be described with its dimensions, its weight, its size, or its colour.

All values in Julia are true objects belonging to a given type (they are individual "instances" of the given type).

Julia types include so-called *primitive types* made of a fixed amount of bits (like all numerical types such as `Int64`, `Float64`, and `Char`) and *composite types* or *structures*, where the object's characteristics are described using multiple fields and a variable number of bits.

Both *structures* and *primitive types* can be user-defined and are hierarchically organized. *Structures* roughly correspond to what are known as *classes* in other languages.

© Antonello Lobianco 2019
A. Lobianco, *Julia Quick Syntax Reference*, https://doi.org/10.1007/978-1-4842-5190-4_4

Two operators are particularly important when working with types:

- The `::` operator is used to constrain an object of being of a given type. For example, `a::B` means "a must be of type B".

- The `<:` operator has a similar meaning, but it's a bit more relaxed in the sense that the object can be of any subtypes of the given type. For example, `A<:B` means "A must be a subtype of B", that is, B is the "parent" type and A is its "child" type.

The primary reason to use these operators is to confirm that the program works the way it is expected. In some cases, providing this extra information can also improve performance.

4.1 Primitive Type Definition

A user-defined primitive type is defined with the keyword `primitive type` and its name and the number of bits it requires:

```
primitive type [name] [bits] end
```

For example:

```
primitive type My10KBBuffer 81920 end
```

❗ A (current) limitation of Julia is that the number of bits must be a multiple of eight below 8388608.

Optionally, a parent type can be specified as follows:

```
primitive type [name] <: [supertype] [bits] end
```

Note that the internal representation of two user-defined types with the same number of bits is exactly the same. The only thing that would change is their names, but that's an important difference: it's the way that functions act when objects of these types are passed as arguments that changes, i.e. it's the use of named types across the program that distinguishes them, rather than their implementation.

4.2 Structure Definition

To define instead a structure, you use the keyword `mutable struct`, give the structure a name, specify the fields, and close the definition with the end keyword:

```
mutable struct MyOwnType                                    JULIA
    field1
    field2::String
    field3::Int64
end
```

Note that while you can optionally define each individual field to be of a given type (e.g., `field3::Int64`), you can't define fields to be subtypes of a given type (e.g., `field3<:Number`). In order to do that, you can use templates in the structure definition:

```
mutable struct MyOwnType{T<:Number}                         JULIA
 field1
 field2::String
 field3::T
end
```

Using templates, the definition of the structure is dynamically created the first time an object whose `field3` is of type T is constructed.

The type with whom you annotate the individual fields can be either a primitive one (like in the previous example) or a reference to another structure (see later for an example).

Note also that with Julia, unlike the other high-level languages (e.g., Python), you can't add or remove fields from a structure after you define it. If you need this functionality, you must use dictionaries instead, but be aware that you will trade this flexibility for worse performance.

Conversely, to gain performance (but again sacrificing flexibility), you can omit the `mutable` keyword in front of `struct`. That way, once an object of that type has been created, its fields can no longer be changed (i.e., structures are immutable by default).

Note that mutable objects—as arrays—also remain mutable in an immutable structure.

4.3 Object Initialization and Usage

After you define your custom type (primitive or structure), you can initialize as many objects as you wish from that type:

```
myObject = MyOwnType("something","something",10)              JULIA
a = myObject.field3 # 10
myObject.field3 = 20 # only if myObject is a mutable struct
```

Note that you have to initialize the object with the values in the same order as was specified in the structure definition. You can't initialize objects by name (rather, consider named tuples for that):

```
MyOwnType(field3 = 10, field1 = "something", field2 =
"something") # Error!
```

To access object fields or change their value, you use, like in most other languages, the `object.field` syntax.

💡 *Functions and Structures*

Note that functions that deal with objects are not defined or declared within the type definition, i.e. they are not associated with one specific type.

This implies that instead of calling an object method, such as `myobj. func(x,y)`, in Julia you would pass the object as a parameter, such as `func(myobj,x,y)`, no matter whether it's as a first, second, or further argument.

4.4 Abstract Types and Inheritance

You can create abstract types using the keyword `abstract type`. Abstract types are not allowed to hold any field, and objects cannot be instantiated from them. Rather, concrete types with fields can be defined as subtypes of them and objects instantiate from them. An issue (see `https://github. com/JuliaLang/julia/issues/4935`) to allow abstract classes to actually hold fields is currently open and may be implemented in the next major release of Julia.

You can create a whole hierarchy of abstract types, although multiple-inheritance (when a type is a subtype of multiple types) is not currently supported:

```
abstract type MyOwnGenericAbstractType end                          JULIA
abstract type MyOwnAbstractType1 <: MyOwnGenericAbstractType end
abstract type MyOwnAbstractType2 <: MyOwnGenericAbstractType end
mutable struct AConcreteType1 <: MyOwnAbstractType1
  f1::Int64
  f2::Int64
end
```

```julia
mutable struct AConcreteType2 <: MyOwnAbstractType1
  f1::Float64
end

mutable struct AConcreteType3 <: MyOwnAbstractType2
  f1::String
end
```

❗ You can't define subtypes of concrete types, only of abstract ones.

Why would you need inheritance if abstract types cannot have fields? It's not to save the time of defining common fields across multiple types at once (you can use *composition* for that, discussed later).

Rather, it is again in the *usage* of types across the program that the definition of their hierarchical structure becomes useful.

Consider the following objects:

```julia
o1 = AConcreteType1(2,10)                                    JULIA
o2 = AConcreteType2(1.5)
o3 = AConcreteType3("aa")
```

By the fact they are all subtypes of MyOwnGenericAbstractType, you can define a function that provides a default implementation for them:

```julia
function foo(a :: MyOwnGenericAbstractType)                  JULIA
  println("Default implementation: $(a.f1)")
end
foo(o1) # Default implementation: 2
foo(o2) # Default implementation: 1.5
foo(o3) # Default implementation: aa
```

You can then decide to provide a function to offer a more specialized implementation for all objects whose type is a subtype of MyOwnAbstractType1:

```julia
function foo(a :: MyOwnAbstractType1)
  println("A more specialised implementation: $(a.f1*4)")
end
foo(o1) # A more specialised implementation: 8
foo(o2) # A more specialised implementation: 6.0
foo(o3) # Default implementation: aa
```

This is possible, thanks to the *multiple dispatch* mechanism discussed in Chapter 3. When multiple *methods* of a function are available to dispatch a function call, Julia will choose the stricter one, i.e., the one defined over the exact parameter's types or their more immediate *supertypes* (parent types):

```julia
function foo(a :: AConcreteType1)
    println("A even more specialised implementation: $(a.f1 +
    a.f2)")
end
foo(o1) # An even more specialized implementation: 12
foo(o2) # A more specialized implementation: 6.0
foo(o3) # Default implementation: aa
```

4.4.1 Implementation of the Object-Oriented Paradigm in Julia

Julia allows both inheritance and composition models, although with different levels of support.

Inheritance is what you just saw, when a hierarchical structure of types is obtained declaring one type as a subtype of another. You also saw the limits of using inheritance in Julia, when only the behavior (and not the structure) can be inherited.

So, how do you implement an object-oriented model in Julia? Well, the preference in the Julia community is to use composition over inheritance.

Composition is when you declare one field of a given type as being an object of another composite type. Through this (referenced) object, you then gain access to the fields of the other type.

Consider the following example, where you first define a generic Person structure and then two more specific Student and Employee structures:

```julia
struct Shoes
    shoesType::String
    colour::String
end

struct Person
  myname::String
  age::Int64
end

struct Student
    p::Person
    school::String
    shoes::Shoes
end

struct Employee
    p::Person
    monthlyIncomes::Float64
    company::String
    shoes::Shoes
end
```

Instead of using inheritance declaring Student and Employee as subtypes of Person, you use composition to assign a field p of type Person to both of them. It is thanks to this field that you do not need to repeat the fields that are common to both.

❗ Types must be defined before they can be used to reference objects. Hence, the Shoes definition must come before the Students and Employee definitions.

You can then create instances of the specialized type, either by creating the referenced object first, or doing that inline in the constructor of the specialized type:

```
gymShoes = Shoes("gym","white")                                    JULIA
proShoes = Shoes("classical","brown")

Marc = Student(Person("Marc",15),"Divine School",gymShoes)
MrBrown = Employee(Person("Brown",45),3200.0,"ABC Corporation
Inc.", proShoes)
```

Finally, you can use multiple dispatch to provide tailored implementation for the specialized types and access referenced objects and/or general fields through a chained use of the dot (.) operator:

```
function printMyActivity(self::Student)                             JULIA
println("Hi! I am $(self.p.myname), I study at $(self.school)
school, and I wear $(self.shoes.colour) shoes")
end

function printMyActivity(self::Employee)
println("Good day. My name is $(self.p.myname), I work at
$(self.company) company and I wear
$(self.shoes.colour) shoes")
end
```

```
printMyActivity(Marc)        # Hi! I am Marc, ...
printMyActivity(MrBrown)     # Good day. My name is MrBrown, ...
```

While using inheritance wisely can in practice suffice for many modeling design situations you may encounter, it is still true that the Julia core language misses some expressiveness in the sense that you cannot directly consider and/or distinguish between different concepts of relations between objects, like specialization (e.g., Person→Student), composition (e.g., Person→Arm), and weak relation (e.g., Person→Shoes).

To address this lack, several third-party packages (not discussed in this book) have been released that improve Julia flexibility in this area, like the SimpleTraits (`https://github.com/mauro3/SimpleTraits.jl`) package to imitate multiple inheritance or `OOPMacro.jl` (`https://github.com/ipod825/OOPMacro.jl`) to automatically copy field declaration from a parent to a child type.

4.5 Some Useful Functions Related to Types

To complete the discussion concerning user-defined types, you may find the following type-related functions useful:

- `supertype(MyType)`: Returns the parent types of a type

- `subtypes(MyType)`: Lists all children of a type

- `fieldnames(MyType)`: Queries all the fields of a structure

- `isa(obj,MyType)`: Checks if `obj` is of type `MyType`

- `typeof(obj)`: Returns the type of `obj`

- `eltype(obj)`: Returns the inner type of a collection, like the inner elements of an array

CHAPTER 5

Input/Output

This chapter includes discussion of the following third-party packages:

Package Name	URL	Version
CSV.jl	https://github.com/JuliaData/CSV.jl	v0.5.11
HTTP.jl	https://github.com/JuliaWeb/HTTP.jl	v0.8.4
XLSX.jl	https://github.com/felipenoris/XLSX.jl	v0.5.4
JSON2.jl	https://github.com/quinnj/JSON2.jl	v0.3.1

Input/Output in Julia is implemented by first choosing the appropriate IOStream object—a file, the user terminal, or a network object—and then applying the desired function over that stream. The general idea is to open the stream, perform the required operations, and then close the stream.

Concerning the user's terminal, Julia provides interaction with it through the built-in stdin and stdout (and eventually stderr) IO streams. These are already opened and are the default streams for input and output operations.

Concerning web resources, you currently still need to use a third-party package. We will consider HTTP.jl (see https://github.com/JuliaWeb/HTTP.jl) in this chapter.

© Antonello Lobianco 2019
A. Lobianco, *Julia Quick Syntax Reference*, https://doi.org/10.1007/978-1-4842-5190-4_5

5.1 Reading (Input)

5.1.1 Reading from the Terminal

- aString = readline(): Read whatever the user types into the terminal until pressing `Enter`.

- anInteger = parse(Int64, readline()): Read a number entered by the user in the terminal.

Let's note that you will rarely want this. Julia is interactive and you can run portions of code, or change the variables directly from the terminal or the editor that is being used.

Still, you can create a Julia script (like in the following example) and run it (with either julia myScript.jl or with include("myScript.jl") from within the Julia prompt). This will mimic the classic "scripts" behavior, where the user runs a script without having full control over it, but only entering the user input that is requested by the script:

```
println("Welcome to a Julia script\n\n")                          JULIA

function getUserInput(T=String,msg="")
  print("$msg ")
  if T == String
      return readline()
  else
    try
      return parse(T,readline())
    catch
      println("Sorry, I could not interpret your answer. Please
      try again")
      getUserInput(T,msg)
    end
  end
end
```

```julia
sentence = getUserInput(String,"Which sentence do you want to
be repeated?");
n = getUserInput(Int64,"How many times do you want it to be
repeated?");
[println(sentence) for i in 1:n]
println("Done!")
```

Note that in this code snippet, function uses runtime exceptions to be sure that the user's response is compatible with the type you want. You'll learn more about runtime exceptions in Chapter 8.

5.1.2 Reading from a File

File reading (and, you will see, writing) is similar in Julia to other languages. You open the file, specify the modality (r, which stands for "reading"), bind the file to an object, operate on the object, and close() it when you are done.

A better alternative is to encapsulate the file operations in a do block that closes the file automatically when the block ends.

- Reading a whole file in a single operation:

```julia
open("afile.txt", "r") do f     # "r" for reading     JULIA
    filecontent = read(f,String) # Note that it can
be used only once. If used a second time, without
reopening the file, read() would return an empty string
    # ... do what you want with the file content ...
end
```

- Reading a file line by line:

```julia
open("afile.txt", "r") do f                           JULIA
    for ln in eachline(f)
        # ... do what you want with the line ...
    end
end
```

This method can be used even with very large files that don't fit into memory.

- Reading a file line by line and keeping track of the line numbers:

```julia
open("afile.txt", "r") do f
   for (i,ln) in enumerate(eachline(f))
     println("$i $ln")
   end
end
```

Importing Data for a Matrix

You can import data from a file to a matrix using readdlm() (in the standard library package DelimitedFiles).

You can skip rows and/or columns using the slice operator and then convert to the desired type:

```
myData = convert(Array{Float64,2},readdlm("myinputfile.
csv",'\t')[2:end,4:end]); # Skip the first 1
row and the first 3 columns
```

Parsing Comma Separated Value (CSV) Files

You can use the read function of the CSV.jl package:

```
myData = CSV.read(file; delim=';', missingstring="NA",
decimal=',')
```

read supports a long list of (mostly self-explanatory) options: delim (use \t for tab-delimited files), quotechar, openquotechar, closequotechar, escapechar, missingstring, dateformat, append, writeheader, header, newline, quotestrings, decimal, header, normalizenames, datarow, skipto, footerskip, limit, transpose, comment, use_mmap, type, types, typemap, pool, categorical, strict, and silencewarnings.

The result is a *DataFrame* (a widely used tabular format of data that you will learn about extensively in Chapter 10). Unless the option header=false has been used, the first row would be interpreted as the header.

The type of each field is auto-recognized using a large number of initial rows, but sometimes this is not enough. For example, in some datasets you may have a field with all missing values and then suddenly non-missing values appear very late in the dataset. Such situations would trigger a TypeError when you would try to populate the field with some values, as it would have been recognised of type Missing.

The trick is to manually specify the column type with the types parameter (a vector or a dictionary, such as types=Dict("myFieldFoo" => Union{Missing,Int64})).

While you used the CSV.jl package here, you may be interested in an alternative, newer package called CSVFiles.jl (see https://github.com/queryverse/CSVFiles.jl), which allows you to access the file directly by URL or if the file is gzipped.

5.1.3 Importing Data from Excel

Using the XLSX package, you can:

- Retrieve a list of the sheet's names: XLSX.
  sheetnames(XLSX.readxlsx("myExcelFile.xlsx"))

- Import all the data from a given sheet: m = XLSX.
  readxlsx("myExcelFile.xlsx")["mySheet"][:]

- Import all the data from a specific interval: m =
  XLSX.readxlsx("myExcelFile.xlsx")["mySheet"]
  ["B3:D6"] or m = XLSX.readdata("myExcelFile.
  xlsx", "mySheet", "B3:D6")

- Import the data to a DataFrame: df =
  DataFrame(XLSX.readtable("myExcelFile.xlsx",
  "mySheet")...)

The last method deserves an explanation. XLSX normally exports data in standard matrix format (`Array{T,2}`). To facilitate the exporting of data in a DataFrame format, another function `XLSX.readtable()` is provided that returns a tuple instead, where the first element is a vector of column data (i.e. `Array{Array{T,1},1}`) and the second element of the tuple is a vector of header names. This format, using the splat constructor to split the tuple into two separate arguments, allows you to match one of the DataFrames constructor methods, namely `DataFrame(data, headers)`.

If you want to import only a range of the sheet to a DataFrame, you can specify in `readtable()` the columns you want to import as the third positional argument (e.g., `B:D`) and the `first_row` from which to import. However, at that point, you are probably better off using a package such as `ExcelFiles.jl` (see `https://github.com/queryverse/ExcelFiles.jl`), which allows directly importing into the `DataFrame` format.

5.1.4 Importing Data from JSON

JavaScript Object Notation (JSON) is a common open standard language-independent data format that uses human-readable text (made of attribute–value pairs) to serialize and transmit data objects.

Here, I present the `JSON2.jl` package (`https://github.com/quinnj/JSON2.jl`). Other common options for reading or exporting data into the JSON format are `JSON.jl` (`https://github.com/JuliaIO/JSON.jl`) and `LazyJSON.jl` (`https://github.com/JuliaCloud/LazyJSON.jl`) (The latter in particular parses the JSON file only when it's needed rather than parsing it all at once.).

Let's assume you have the following JSON data available in Julia as strings, representing an extract of the Nottingham Forest:

```
jsonString="""                                          JULIA
{
    "species": "Oak",
   "latitude": 53.204199,
```

```
    "longitude": -1.072787,
        "trees": [
                    {
                        "vol": 23.54,
                        "id": 1
                    },
                    {
                        "vol": 12.25,
                         "id": 2
                    }
                    ]
}
"""
```

There are two ways to import this data in your program. The first way is to use JSON2.read(jsonString, T), that is, to import the data as a specific type (that must have already being defined):

```
struct ForestStand                                           JULIA
    sp::String
    lat::Float64
    long::Float64
    trees::Array{Dict{String,Float64},1}
end
nottFor = JSON2.read(jsonString, ForestStand)
```

nottFor will then be a ForestStand object. Note that the names of the fields defined in the structure and in the imported JSON data don't need to match. What matters is the order and types of the fields.

The second way is to not specify any type for the imported object, as follows:

```
nottFor2 = JSON2.read(jsonString)
```

nottFor2 will then be a NamedTuple whose fields can be accessed with the normal dot notation, e.g., nottFor2.trees[1].vol.

5.1.5 Accessing Web Resources (HTTP)

Web resources can be accessed through the Hypertext Transfer Protocol (HTTP) using the third-party package called HTTP.jl (https://github.com/JuliaWeb/HTTP.jl)[1]. The HTTP.jl package is quite complete and includes a server side. Here, you just see the basic way to access (read) a web resource.

First, you could use HTTP.jl to provide a stream and operate over it, in a similar way to how you did with a local file:

```julia
HTTP.open("GET", "https://julialang.org/") do io
# Note the SSLsupport
    while !eof(io)
        println(String(readavailable(io)))
    end
end
```

Alternatively, you can use more concise syntax with the get method:

```julia
res = HTTP.get("https://julialang.org/" )
println(String(res.body))
```

If all you need to do is read from a web resource the second method is probably simpler. The first method has the advantage that you can open other streams at the same time and redirect the content of one stream to another. For example, you can redirect the stream from a remote video to the local player.

---

[1] A reflection is under way in the Julia community to move HTTP.jl to the Julia standard library.

Finally, you can use CSV.jl and HTTP.jl together to access and parse remote CSV files:

```julia
using HTTP, CSV
resp = HTTP.request("GET", "https://cohesiondata.ec.europa.eu/
resource/2q3n-nr7n.csv")
df = CSV.read(IOBuffer(String(resp.body)))
```

5.2 Writing (Output)

5.2.1 Writing to the Terminal

To write to the terminal, you can use the write(IO,T) function or print(IO,T), using the built-in stdout (default for print) or stderr streams.

- write(stdout, "Hello World");: Print the "Hello World" string to the terminal. Note the semicolon ; at the end. Without it, the write method would also print the size (of characters) of the string printed.

- print("Hello World") ;

- println("Hello World"): Call print("Hello World") and add a newline.

🛈 *write versus print*

The main difference between the two functions is that while write outputs the raw bytes of the object, print outputs a text representation of it. This representation is already defined for all the built-in types, and a default is provided for custom types (in contrast with `write()`, which would throw a `MethodError`). This default can be modified by overriding `print()`:

```julia
import Base.print # Needed when we want to
override the print function                      JULIA
struct aCustomType
    x::Int64
    y::Float64
    z::String
end
foo = aCustomType(1,2,"MyObj")
print(foo) # Output: aCustomType(1, 2.0, "MyObj")
println(foo) # Output: aCustomType(1, 2.0, "MyObj")
+ NewLine
write(stdout,foo) # Output: MethodError
function print(io::IO, c::aCustomType)
    print("$(c.z): x: $(c.x), y: $(c.y)")
end
print(foo) # Output: MyObj: x: 1, y: 2.0
println(foo) # Output: MyObj: x: 1, y: 2.0 +
NewLine (no need to override also `println()``)
```

5.2.2 Writing to a File

Writing to a file is similar to reading a file, but with the modality w (overwrite the file) or a (append to the existing data) instead of r in the open() function call. You must also, of course, use an output function as write or print over the IO object:

```julia
open("afile.txt", "w") do f # "w" for writing
  write(f, "First line\n") # \n for newline
  println(f, "Second line") # Newline automatically added by
  println
end
```

If you prefer to use the pattern open > work on file > close instead of using the do block, be aware that the operations will be flushed on the file on-disk only when the IOStream is closed or when a certain buffer limit is reached.

5.2.3 Exporting to CSV

Exporting to a CSV file is undoubtedly simpler than reading a CSV file, as you don't need to account for the myriad of possible CSV options and are in control of which format to give to the file.

You can simply loop over the rows/columns of data and manually write down the data and the delimiters. Or, you can use the CSV.write() functions, which does that for you:

```julia
CSV.write("myOutputFile.csv", myData, delim=';', decimal='.',
missingstring="NA")
```

This works out of the box when the data to export is a DataFrame, but for a standard matrix (Array{T,2}), you need first to transform it into a so-called MatrixTable using the function Tables.table(Matrix) from the Tables package (see https://github.com/JuliaData/Tables.jl):

```julia
CSV.write("myOutputFile.csv", Tables.table(myMatrix),
delim=';', decimal='.', missingstring="NA", header=["field1",
"field2","field3"])
```

write supports the following options: delim, quotechar, openquotechar, closequotechar, escapechar, missingstring, dateformat, append, writeheader, header, newline, quotestrings, and decimal.

5.2.4 Exporting Data to Excel

Writing to Excel follows the common pattern of opening the file and operating on it within a do block.

```
XLSX.openxlsx("myExcelFile.xlsx", mode="w") do xf
# w to write (or overwrite) the file                    JULIA
    sheet1 = xf[1] # One sheet is created by default
    XLSX.rename!(sheet1, "new sheet 1")
    sheet2 = XLSX.addsheet!(xf, "new sheet 2") # We can add
    further sheets if needed
    sheet1["A1"] = "Hello world!"
    sheet2["B2"] = [ 1 2 3 ; 4 5 6 ; 7 8 9 ] # Top-right cell
    to anchor the matrix
end
```

This code creates a new file (if a file with the same name already exists, it will be overwritten). In order to "append" the data to an existing file without rewriting it, use mode="rw" instead:

```
XLSX.openxlsx("myExcelFile.xlsx", mode="rw") do xf
# rw to append to an existing file instead              JULIA
    sheet1 = xf[1] # One sheet is created by default
    sheet2 = xf[2]
    sheet3 = XLSX.addsheet!(xf, "new sheet 3") # We can add
    further sheets if needed
    sheet1["A2"] = "Hello world again!"
    sheet3["B2"] = [ 10 20 30 ; 40 50 60 ; 70 80 90 ] # Top-
    right cell to anchor the matrix
end
```

Individual cells can accept Missing, Bool, Float64, Int64, Date, DateTime, Time, and String values.

As shown in the last line of this example, arrays (both `Array{T,1}` and `Array{T,2}`) are automatically broadcast to individual cells. This is not true for other data structures, like DataFrames. In the same way as XLSX offers a specific function to facilitate importing to a DataFrame, it also offers a function to facilitate its exports. This function is `writetable`, where the data to be exported is given by a tuple (`Array{Array{T,1},1}`, `Array{String}`), that is columns of data and then field names. This format can be easily obtainable from a DataFrame:

```
XLSX.writetable("myNewExcelFile.xlsx", sheet1=( [ [1, 2, 3],
[4,5,6], [7,8,9]], ["f1","f2","f3"] ), sheet2= (collect
(DataFrames.eachcol(myDf)), DataFrames.names(myDf) ))
```

Note that multiple sheets can be written at once, but the destination file must be new (or an error will be raised).

5.2.5 Exporting Data to JSON

Using the `JSON2.jl` package again (`https://github.com/quinnj/JSON2.jl`) and reusing the `nottFor` object defined in the "Importing Data from JSON" section, you can export any object to JSON simply by using `jsonString = JSON2.write(nottFor)`.

The returned string is a valid JSON string, but it's written as a single line and quoted.

To display a human-readable version of it, use the `@pretty` macro provided by the same package: `@pretty jsonString`.

5.3 Other Specialized IO

All major data-storage formats have at least one Julia package to allow you to interact with them. Here is a partial list (with my suggestions listed first):

- XML: `EzXML.jl` (see `https://github.com/bicycle1885/EzXML.jl`) and `LightXML.jl` (see `https://github.com/JuliaIO/LightXML.jl`)

- HTML (web scraping): `Gumbo.jl` (see `https://github.com/JuliaWeb/Gumbo.jl`) and `Cascadia.jl` (see `https://github.com/Algocircle/Cascadia.jl`)

- OpenDocument spreadsheet format (ODS): `OdsIO.jl` (see `https://github.com/sylvaticus/OdsIO.jl`)

- HDF5: `HDF5.jl` (see `https://github.com/JuliaIO/HDF5.jl`)

CHAPTER 6

Metaprogramming and Macros

Metaprogramming is the technique that programmers use in order to write code (computer instructions) that, instead of being directly evaluated and executed, produces different code that is in turn evaluated and run by the machine.

For example, the following C++ macro is, in a broad sense, metaprogramming:

```cpp
#define LOOPS(a,b,c) \
  for (uint i=0; i<a; i++){ \
    for (uint j=0;j<b; j++){ \
      for (uint z=0;z<c; z++){ \
```

Here, the programmer, instead of writing the three for loops (supposedly in many places of her program), can just write LOOPS(1,2,3) and this will be expanded and substituted for the loops in the actual code before it's run.

In this case, as in most programming languages, the "metaprogramming" usage remains fairly basic. First, it requires a different building step than the main one (in C++, this is done by the preprocessor). Second, and perhaps most important, the substitution is done at the level of a textual "search and replace," without any syntactical interpretation of the code as computer instructions.

© Antonello Lobianco 2019
A. Lobianco, *Julia Quick Syntax Reference*, https://doi.org/10.1007/978-1-4842-5190-4_6

In Julia and a few other languages (notably those derived or inspired by the Lisp programming language), the code written by the programmer is parsed and translated to computer instructions in what is known as the Abstract Syntax Tree (or AST). At this point, the code is in the form of a data structure that can be traversed, created, and manipulated from within the language (without the need for a third-party tool or a separate building step). That means that the "code substitution" (later called a *macro*) can now be much more expressive and powerful, as you can directly manipulate computer instructions instead of text.

This chapter introduces the concepts of *symbols* and *expressions*, and then explains how macros work.

6.1 Symbols

Symbols are intimately bound with the ability of Julia to represent the language's code as a data structure in the language itself. A *symbol* is indeed a way to refer to a data object and still keep it in an unevaluated form. For example, when dealing with a variable, you can use a symbol to refer to the actual identifiers and not to the variable's value. Symbols can also refer to operators and any other parts of the (parsed) computer instructions. For example, `:myVar`, `:+`, and `:call` are all valid symbols.

To form a symbol, use the colon `:` prefix operator or the `symbol()` function. For example:

```
a = :foo10 is equal to a=Symbol("foo10")
```

The `symbol()` function can optionally concatenate its arguments to form the symbol, as follows:

```
a=Symbol("foo",10)
```

To convert a symbol back to a string, use `string(mysymbol)`.

6.2 Expressions

Expressions are unevaluated computer instructions. In Julia they too
are objects. Specifically, they are instances of the expr type, whose fields
are head, defining the kind of expression, and args, defining the array of
elements. These can be symbols, primitive (not modifiable) values such as
strings and numbers, or other sub-expressions (from which the tree data
structure is obtained).

Being objects, expressions are first-class citizens in Julia. They support
all the operations generally available to other entities, as being passed as
an argument, returned from a function, modified, or assigned to a variable.

Expressions are normally created by parsing the computer code and
their ultimate destiny is to be evaluated.

For example, the string "b = a+1" when parsed becomes an
expression. In the AST, this is seen as follows:

```
julia> expr = Meta.parse("b = a+1 # This is a comment")    JULIA
julia> typeof(expr)
Expr
julia> dump(expr)
Expr
  head: Symbol =
  args: Array{Any}((2,))
    1: Symbol b
    2: Expr
      head: Symbol call
      args: Array{Any}((3,))
        1: Symbol +
        2: Symbol a
        3: Int64 1
```

In the first line of the script, you start creating an expression object by parsing the string "b=a+1 # This is a comment" and saving it in the expr variable. You then dump it in order to examine its internals.

Note that the comments have been stripped out and that this simple expression is interpreted as a branch. The master is the equals sign and the two arguments are the variable b and a lower-level expression, which itself is a call to the + operator with the parameters a and 1.

This shows you that internally, b = a+1 is seen as b = +(a,1), i.e., the plus operator is acting like a normal function that takes parameters.

6.2.1 Creating Expressions

There are many ways to create expressions, discussed next.

Parse a String

The first way to create an expression is what you just saw—by parsing a string with meta.parse(). This is what Julia uses when parsing a .jl script or the REPL input.

Colon Prefix Operator

Expressions can be created using the same colon : prefix operator you saw for individual symbols, this time applied to a whole expression (given in brackets):

```
expr = :(a+1)
```

Quote Block

An alternative to the :([...]) operator is the quote [...] end block:

```
expr = quote                                         JULIA
          b = a+1
       end
```

Use the Exp Constructor with a Tree

An expression can be also directly constructed from a given tree. So this:

```
expr2 = Expr(:(=), :b, Expr(:call,:+,:a,1))
```

is equivalent to this:

```
 expr = Meta.parse("b = a+1") or expr = :(b = a+1)
```

6.2.2 Evaluating Symbols and Expressions

To evaluate an expression or a symbol, use the eval() function:

```
expr = Meta.parse("3+2")                                    JULIA
eval(expr) # 5
```

> ❗ The evaluation happens at the global scope, even if the eval()
> call is done within a function. That is, the expression being evaluated
> will have access to the global variables but not to the local ones.

This code is trivial, as it only contains literals, i.e., immutable values known at compile time. But what happens when the expression contains variables? Will the computation consider the value bound to the variable at the time the expression is composed or when it is evaluated?

It ends up that you can choose this when you compose the expression, as shown in the following snippet:

```
a = 1                                                       JULIA
expr1 = Meta.parse("$a + b + 3")
expr2 = :($a + b + 3)              # Equiv. to expr1
expr3 = quote $a + b + 3 end       # Equiv. to expr1
expr4 = Expr(:call, :+, a, :b, 3)  # Equiv. to expr1
```

```
eval(expr1)     # UndefVarError: b not defined
b = 10
eval(expr1)     # 14
a = 100
eval(expr1)     # Still 14
b = 100
eval(expr1)     # 104
```

In the first line, you define the variable a and assign it a starting value. In expressions 1 to 3, you use the dollar $ operator to interpolate ("unquote") the variable a. This indicates that you want to store its content in the expression, not the variable itself. This is why a has to exist beforehand.

Conversely, with b, you indicate that it is the identifier b and not the value that is stored in the expression. This variable doesn't even need to be defined at this point.

Note that when you use the Expr constructor method to define expr4, you deal directly with unquoted entities. When you indicate a, you hence mean *the value bound to it*, and to indicate an identifier instead, you need to use a symbol (such as :b).

When you try to evaluate the expression, at this point Julia tries to look up the b symbol, and not finding it, throws an error. You must define and assign a value to b before the expression can be evaluated.

Finally, note that whatever happens to the variable a has no effect on the evaluations of the expression, but modifying the value of b affects it.

ⓘ *A Loose Comparison to C/C++ Pointers*

In C/C++, the & operator, when applied to a variable, returns its memory address. Conversely, the * operator, when applied to a memory address, returns the value of the variable stored at that address.

In Julia, you can think of the variable name as the address itself. That means :var (like &var) returns the variable name, while $var (like *var) returns the value bound to that variable.

6.3 Macros

The possibility to represent code as expressions is at the heart of macros' power. Macros in Julia take one or more input expressions (and optionally literals and symbols) and return a modified expressions at parse time. Contrast this with normal functions that, at runtime, take input values (arguments) and return a computed value.

Macros move the computation from the runtime to the compile-time, as the expression is generated and compiled directly rather than requiring a runtime eval call like a function would.

6.3.1 Macro Definition

Defining a macro is similar to defining a normal function. The differences are that you use the macro keyword in place of the function keyword and you compose the expression that the macro should return. For example, using a quote block where the variables used as parameters are unquoted using the dollar symbol. That's make sense, as the parameters of the macro are expressions, and so using the dollar symbol for the variable bound to them means injecting their content (an expression) within the broader context of the whole expression defined in the quote block.

Let's look at an example:

```
macro customLoop(controlExpr,workExpr)                    JULIA
  return quote
    for i in $controlExpr
      $workExpr
```

```
      end
   end
end
```

Here you use a macro to define a generic loop, which is given by both the control you apply to the `for` loop and the expression(s) to be evaluated inside the loop.

6.3.2 Macro Invocation

Once a macro is defined, you can call it using its name, prefixed with the at symbol (@). The macro is then followed by the expressions that it will accept in the same row, separated by spaces (use the `begin` block syntax to group expressions together).

For example:

```
a = 5                                                      JULIA
@customLoop 1:4 println(i)
@customLoop 1:a println(i)
@customLoop 1:a if i > 3 println(i) end
@customLoop ["apple", "orange", "banana"] println(i)
@customLoop ["apple", "orange", "banana"] begin print("i: ");
println(i) end
```

You can see how the "expanded" macro will look using, well, another macro, called @macroexpand:

```
julia> @macroexpand @customLoop 1:4 println(i)          JULIA
quote
    #= /path/to/source/file/metaprogramming.jl:65 =#
    for #92#i = 1:4
        #= /path/to/source/file/metaprogramming.jl:66 =#
```

```
      (Main.println)(#92#i)
    end
end
```

Note that a macro doesn't create a new scope, and variables declared or assigned within the macro may collide with variables in the scope at the point where the macro is actually called.

6.3.3 String Macros

Finally, a convenient type of macro (whose long but technically correct name is "non-standard string literals") allows developers to perform compile-time custom operations on text entered as xxx"...text...". This is a custom prefix that's immediately attached to the text to be processed and entered as a string (tripled quotas, such as xxx"""...multi-line text...""" can be used as well).

For example, the following macro defines a custom eight-column display of the entered text:

```
macro print8_str(mystr)                                    JULIA
  limits = collect(1:8:length(mystr))
  for (i,j) in enumerate(limits)
    st = j
    en = i==length(limits) ? length(mystr) : j+7
    println(mystr[st:en])
  end
end
```

These special macros, whose names end with _str, can then be called using a non-standard string literal, where the prefix matches the macro name without the _str part:

```julia
julia> print8"123456789012345678"                          JULIA
12345678
90123456
78
```

You will see applications of these macros in the next chapter, "Interfacing Julia with Other Languages," where string macros will be used to process text representing code in a foreign language (C++, Python, or R).

CHAPTER 7

Interfacing Julia with Other Languages

This chapter includes a discussion of the following third-party packages:

Package Name	URL	Version
Cxx.jl	https://github.com/JuliaInterop/Cxx.jl	v0.3.2
PyCall.jl	https://github.com/JuliaPy/PyCall.jl	v1.91.2
PyJulia	https://github.com/JuliaPy/pyjulia	v0.4.1
RCall.jl	https://github.com/JuliaInterop/RCall.jl	v0.13.3
JuliaCall	https://github.com/Non-Contradiction/JuliaCall	v0.16.5

Julia is a relatively new language (the first public release is dated February 2012), which means there is obvious concern regarding the availability of packages that implement its functionalities.

To address this concern, one specific area of Julia development addresses its capacity to interface with code written in other languages, and in particular to use the huge number of libraries available in those languages.

At its core, this resulted in the capacity of the Julia language to natively call C and FORTRAN libraries. This in turn allowed Julia developers to write packages to interface with the most common programming languages.

© Antonello Lobianco 2019
A. Lobianco, *Julia Quick Syntax Reference*, https://doi.org/10.1007/978-1-4842-5190-4_7

I will discuss some of them in this chapter. Finally, this allowed higher-level packages to leverage these "language-interface" packages to easily wrap existing libraries, written in other languages, and present an interface to them for Julia programmers.

To sum up, if you need a common functionality, it is highly likely that you will find a package that either (a) implements the functionality directly in Julia (such as DataFrames (`https://github.com/JuliaData/DataFrames.jl`)) or (b) wraps an existing library in another language (for the same functionality, such as Pandas—`https://github.com/JuliaPy/Pandas.jl`).

I will discuss some of these packages in the second part of the book. The rest of this chapter will show how to explicitly link Julia with other languages, either by calling functions implemented in other languages or by calling Julia functions from other languages.

ℹ When it would be ambiguous (such as when using prompts of different languages), I include the prompt symbols in the code snippets.

7.1 Julia ⇄ C

As stated, calling C code is native to the language and doesn't require any third-party code.

Let's first build a C library. I show this in Linux using the GCC compiler. The procedure in other environments is similar but not necessarily identical.

myclib.h:

```
extern int get2 ();                                              C
extern double sumMyArgs (float i, float j);
```

myclib.c:

```c
int get2 (){                                                    C
 return 2;
}
double sumMyArgs (float i, float j){
 return i+j;
}
```

Note that you need to define the function you want to use in Julia as extern in the C header file.

You can then compile the shared library with gcc, a C compiler:

```
gcc -o myclib.o -c myclib.c
```

```
gcc -shared -o libmyclib.so myclib.o -lm -fPIC
```

You are now ready to use the C library in Julia:

```julia
const myclib = joinpath(@__DIR__, "libmyclib.so")              JULIA
a = ccall((:get2,myclib), Int32, ())
b = ccall((:sumMyArgs,myclib), Float64, (Float32,Float32), 2.5, 1.5)
```

The ccall function takes as the *first* argument a tuple (function name, library path), where the library path must be expressed in terms of a full path, unless the library is in the search path of the OS. If it isn't, and you still want to express it relative to the file you are using, you can use the @__DIR__ macro, which expands to the absolute path of the directory of the file. The variable hosting the full path of the library must be set constant, and the tuple acting as the first argument of ccall must be *literal*.

The *second* argument of ccall is the return type of the function. Note that while C int maps to Julia Int32 or Int64, C float maps to Julia Float32 and C double maps to Julia Float64. In this example, you could have instead used the corresponding Julia type aliases Cint, Cfloat, and Cdouble (within others) in order to avoid memorizing the mapping.

The *third* argument is a tuple of the types of the arguments expected by the C function. If there is only one argument, it must still be expressed as a tuple, e.g. (Float64,).

Finally, the remaining arguments are passed to the C function.

I have just scratched the surface here. Linking C (or FORTRAN) code can become pretty complex in real-world situations, and consulting the official documentation (see https://docs.julialang.org/en/v1/manual/calling-c-and-fortran-code/) can prove indispensable.

7.2 Julia ⇄ C++

As with C, C++ workflow is partially environment-dependent. This section uses Cxx.jl under Linux, although Cxx.jl also has experimental support for Windows[1].

Its main advantage over other C++ wrap modules (notably, CxxWrap.jl at https://github.com/JuliaInterop/CxxWrap.jl) is that it allows you to work on C++ code in multiple ways depending on the workflow that is required.

7.2.1 Interactive C++ Prompt

If you type < in the Julia REPL (after you type using Cxx), you obtain an interactive C++ prompt, as shown in Figure 7-1.

---

[1]Cxx is compatible only with Julia 1.1 or 1.2 (although compatibility with Julia 1.3 is expected soon).

```
Documentation: https://docs.julialang.org

Type "?" for help, "]?" for Pkg help.

Version 1.1.0 (2019-01-21)
Official https://julialang.org/ release

julia> using Cxx

C++ > #include <iostream>;

julia> println("Back to Julia (using the Backspace)")
Back to Julia (using the Backspace)

C++ > std::cout << "Hello, I'm typing C++ code interactivly in the Julia prompt ! (with an error)" << endl;
In file included from /Cxx.cpp:1:
REPL:1:97: error: use of undeclared identifier 'endl'; did you mean 'std::endl'?
std::cout << "Hello, I'm typing C++ code interactivly in the Julia prompt ! (with an error)" << endl
                                                                                                ^~~~

/usr/lib/gcc/x86_64-linux-gnu/7.4.0/../../../../include/c++/7.4.0/ostream:590:5: note: 'std::endl' declared here
    endl(basic_ostream<_CharT, _Traits>& __os)
    ^

Hello, I'm typing C++ code interactivly in the Julia prompt ! (with an error)

C++ > std::cout << "Hello, I'm typing C++ code interactivly in the Julia prompt ! (without errors)" << std::endl;

Hello, I'm typing C++ code interactivly in the Julia prompt ! (without errors)
```

Figure 7-1. *C++ interactive prompt at the Julia REPL*

Note in Figure 7-1 how the code compilation you typed is done just-in-time, as with Julia itself, so any compilation errors are reported straight away.

7.2.2 Embed C++ Code in a Julia Program

In addition to the REPL prompt, you can use C++ code in your Julia program without ever leaving the main Julia environment. Let's start with a simple example:

```julia
using Cxx                                                    JULIA

# Define the C++ function and compile it
cxx"""
#include<iostream>
void myCppFunction() {
    int a = 10;
    std::cout << "Printing " << a << std::endl;
}
```

```
"""
# Execute it
icxx"myCppFunction();" # Return "Printing 10"
# OR
# Convert the C++ function to a Julia function
myJuliaFunction() = @cxx myCppFunction()
# Run the function
myJuliaFunction() # Return "Printing 10"
```

The workflow is straight-forward: You first embed the C++ code with the cxx"..." string macro. You are then ready to "use" the functions defined in cxx"..." either by calling them directly with C++ code embedded in icxx"..." or by converting them in a Julia function with the @cxx macro and using Julia code to call the (Julia) function.

This example doesn't imply any transfer of data between Julia and C++. If you want to embed C++, most likely it is in order to pass some data to C++ and retrieve the output to use in your Julia program.

The following example shows how data transfer between Julia and C++ (both ways) is handled automatically for elementary types:

```
using Cxx                                          JULIA
cxx"""
#include<iostream>
int myCppFunction2(int arg1) {
    return arg1+1;
}
"""
juliaValue = icxx"myCppFunction2(9);" # 10
# OR
myJuliaFunction2(arg1) = @cxx myCppFunction2(arg1)
juliaValue = myJuliaFunction2(99) # 100
```

However, when the transfer involves complex data structures (like arrays or, as in the following example, arrays of arrays), things become more complex. The Julia types have to be converted into the C++ types and vice versa:

```
using Cxx                                                    JULIA
cxx"""
#include <vector>
using namespace std;
vector<double> rowAverages (vector< vector<double> > rows) {
  // Compute average of each row..
  vector <double> averages;
  for (int r = 0; r < rows.size(); r++){
    double rsum = 0.0;
    double ncols= rows[r].size();
    for (int c = 0; c< rows[r].size(); c++){
       rsum += rows[r][c];
    }
    averages.push_back(rsum/ncols);
  }
  return averages;
}
"""
rows_julia = [[1.5,2.0,2.5],[4,5.5,6,6.5,8]]
rows_cpp   = convert(cxxt"vector< vector< double > > ", rows_julia)
rows_avgs  = collect(icxx"rowAverages($rows_cpp);")
# OR
rowAverages(rows) = @cxx rowAverages(rows)
rows_avgs = collect(rowAverages(rows_cpp))
```

The conversion from Julia data to C++ data (to be used as arguments of the C++ function) is done using the familiar `convert(T,source)` function, but where T is given by the expression returned by `cxxt"[Cpp type]"`.

The converted data can then be used in the C++ function call, with the note that if the direct call form of `icxx"` is used, this has to be interpolated using the dollar $ operator.

Finally, `collect` is used to copy data from the C++ structures returned by the C++ function to a Julia structure (you can avoid copying by using pointers).

7.2.3 Load a C++ Library

The third way to use C++ code is to load a C++ library and then call the functions defined in that library directly. The advantage to this approach is that the library doesn't need to be aware that it will be used in Julia (i.e., no special Julia headers or libraries are required at the compile time of the C++ library), as the wrap is done entirely in Julia. That means preexisting libraries can be reused.

Let's reuse the last example, but this time you will compile it in a shared library.

`mycpplib.cpp`:

```cpp
#include <vector>
#include "mycpplib.h"
using namespace std;
vector<double> rowAverages (vector< vector<double> > rows) {
  // Compute average of each row.
  vector <double> averages;
  for (int r = 0; r < rows.size(); r++){
    double rsum = 0.0;
    double ncols= rows[r].size();
```

```cpp
  for (int c = 0; c< rows[r].size(); c++){
    rsum += rows[r][c];
  }
  averages.push_back(rsum/ncols);
}
return averages;
}
```

mycpplib.h:

```cpp
#include <vector>
std::vector<double> rowAverages (std::vector<
std::vector<double> > rows);
```

You can then compile the shared library with g++, a C++ compiler:

```
g++ -shared -fPIC -o libmycpplib.so mycpplib.cpp
```

You are now ready to use it in Julia:

```julia
using Cxx
using Libdl

const path_to_lib = pwd()
addHeaderDir(path_to_lib, kind=C_System)
cxxinclude("mycpplib.h")
Libdl.dlopen(joinpath(path_to_lib, "libmycpplib.so"),
Libdl.RTLD_GLOBAL)

rows_julia = [[1.5,2.0,2.5],[4,5.5,6,6.5,8]]
rows_cpp = convert(cxxt"std::vector< std::vector< double > >",
rows_julia)
rows_avgs = collect(icxx"rowAverages($rows_cpp);")
```

The only new steps here are as follows:

- You need to explicitly load the standard lib `Libdl`

- You need to add the C++ headers

- You need to open the shared library

You are now able to use the functions defined in the library as if you embedded the C++ code in the Julia script.

7.3 Julia ⇌ Python

The "standard" way to call Python code in Julia is to use the PyCall (`https://github.com/JuliaPy/PyCall.jl`) package. Some of its nice features include the following:

- It can automatically download and install a local copy of Python, private to Julia, in order to avoid messing with version dependency from the "main" Python installation and provide a consistent environment within Linux, Windows, and MacOS.

- It provides automatic conversion between Julia and Python types.

- It is very simple to use.

Concerning the first point, PyCall by default installs the "private" Python environment in Windows and MacOS while it will use the system default Python environment in Linux.

You can override this behavior with (from the Julia prompt) `ENV["PYTHON"]="blank or /path/to/python"; using Pkg; Pkg.build("PyCall");` where, if the environmental variable is empty, PyCall will install the "private" version of Python.

Given the vast number of Python libraries, it is no wonder that PyCall is one of the most common Julia packages.

7.3.1 Embed Python Code in a Julia Program

Embedding Python code in a Julia program is similar to what we saw with C++, except that you don't need (for the most part) to worry about transforming data. You define and call the Python functions with py"..." and, in the function call, you can use your Julia data directly:

```julia
using PyCall                                        JULIA
py"""
def sumMyArgs (i, j):
  return i+j
def getNElement (n):
  a = [0,1,2,3,4,5,6,7,8,9]
  return a[n]
"""
a = py"sumMyArgs"(3,4)          # 7
b = py"sumMyArgs"([3,4],[5,6])  # [8,10]
typeof(b)                       # Array{Int64,1}
c = py"sumMyArgs"([3,4],5)      # [8,9]
d = py"getNElement"(1)          # 1
```

Note that you don't need to convert complex data like arrays, and the results are converted to Julia types. Type conversion is automatic for numeric, Boolean, string, IO stream, date/period, and function types, along with tuples, arrays/lists, and dictionaries of these types. Other types are converted to the generic PyObject type.

Note in the last line of the previous example that PyCall doesn't attempt index conversion (Python arrays are zero-based while Julia arrays are one-based.). Calling the Python getNElement() function with 1 as an argument will retrieve what *in Python* is the first element of the array.

7.3.2 Use Python Libraries

Using a Python library is straightforward as well, as shown in the following example, which uses the ezodf module (https://github.com/T0ha/ezodf) to create an OpenDocument spreadsheet (A wrapper of ezodf for ODS documents that internally use PyCall already exists. See OdsIO at https://github.com/sylvaticus/OdsIO.jl.).

Before attempting to replicate the following code, be sure that the ezodf module is available in the Python environment you are using in Julia. If this is an independent environment, just follow the Python way to install packages (e.g., with pip).

If you are using the "private" Conda environment, you can use the Conda.jl package (https://github.com/JuliaPy/Conda.jl) and type using Conda; Conda.add_channel("conda-forge"); Conda.add("ezodf").

```julia
const ez = pyimport("ezodf") # Equiv. of Python
`import ezodf as ez`
destDoc = ez.newdoc(doctype="ods", filename="anOdsSheet.ods")
sheet = ez.Sheet("Sheet1", size=(10, 10))
destDoc.sheets.append(sheet)
dcell1 = get(sheet,(2,3)) # Equiv. of Python `dcell1 =
sheet[(2,3)]`. This is cell "D3" !
dcell1.set_value("Hello")
get(sheet,"A9").set_value(10.5) # Equiv. of Python
`sheet['A9'].set_value(10.5)`
destDoc.backup = false
destDoc.save()
```

Using the module in Julia follows the Python API with few syntax differences. The module is imported and assigned to a shorter alias, ez.

You can then directly call its functions with the usual Python syntax, module.function().

The doc object returned by newdoc is a generic PyObject type. You can then access its attributes and methods with myPyObject.attribute and myPyObject.method(), respectively. When you can't directly access some indexed values, such as sheet[(2,3)] (where the index is a tuple), you can invoke the get(object,key) function instead. Finally, note again that index conversion is *not* automatically implemented: when asking for get(sheet,(2,3)) these are interpreted as *Python*-based indexes, and cell D3 of the spreadsheet is returned, not B2.

7.3.3 PyJulia: Using Julia in Python

The other way around, embedding Julia code in a Python script or terminal, is equally of interest, as in many cases it provides substantial performance gains for Python programmers, and it may be easier than embedding C or C++ code.

This section shows how to achieve it using the PyJulia Python package (https://github.com/JuliaPy/pyjulia), a Python interface to Julia, with the warning that, at time of this writing, it is not as polished, simple, and stable as PyCall, the Julia interface to Python.

Installation

Before installing PyJulia, be sure that the PyCall module is installed in Julia and that it is using the same Python version as the one from which you want to embed the Julia code (eventually, run ENV["PYTHON"]="/path/to/python"; using Pkg; Pkg.build("PyCall"); from Julia to change its underlying Python interpreter).

At the moment, only the pip package manager is supported in Python to install the PyJulia package (conda support should come soon).

Notice that the name of the package in `pip` is `julia`, not `PyJulia`:

```python
$ python3 -m pip install --user julia
>>> import julia
>>> julia.install()
```

If you have multiple Julia versions, you can specify the one to use in Python by passing `julia="/path/to/julia/binary/executable"` (e.g., `julia = "/home/myUser/lib/julia-1.1.0/bin/julia"`) to the `install()` function.

Usage

To obtain an interface to Julia, just run:

```python
>>> import julia;
>>> jl = julia.Julia(compiled_modules=False)
```

The `compiled_module=False` line in the Julia constructor is a workaround to the common situation when the Python interpreter is statically linked to `libpython`, but it will slow down interactive experience, as it will disable Julia packages pre-compilation. Every time you use a module for the first time, it needs to be compiled first. Other, more efficient but also more complicated, workarounds are given in the package documentation, under the Troubleshooting section (see `https://pyjulia.readthedocs.io/en/stable/troubleshooting.html`).

You can now access Julia in multiple ways.

You may want to define all your functions in a Julia script and "include" it. Let's assume `juliaScript.jl` is made of the following Julia code:

```julia
function helloWorld()
    println("Hello World!")
end
function sumMyArgs(i, j)
  return i+j
```

```
end
function getNElement(n)
  a = [0,1,2,3,4,5,6,7,8,9]
  return a[n]
end
```

You can access its functions in Python with:

```
>>>> jl.include("juliaScript.jl")                          PYTHON
>>>> jl.helloWorld() # Prints `Hello World!`
>>>> a = jl.sumMyArgs([1,2,3],[4,5,6]) # Returns `array([5, 7, 9],
dtype=int64)`
>>>> b = jl.getNElement(1) # Returns `0`, the "first" element
for Julia
```

As with calling Python from Julia, you can pass to the functions and retrieve complex data types without worrying too much about the conversion. Note that you now get the Julia way of indexing (one-based).

You can otherwise embed Julia code directly into Python using the Julia eval() function:

```
>>> jl.eval("""                                            PYTHON
... function funnyProd(is, js)
...     prod = 0
...     for i in 1:is
...        for j in 1:js
...           prod += 1
...        end
...     end
...     return prod
... end
... """)
```

You can then call this function in Python as jl.funnyProd(2,3).

What if, instead, you wanted to run the function in broadcasted mode, i.e. apply the function for each element of a given array? In Julia, you could use the dot notation, e.g. `funnyProd.([2,3],[4,5])` (this would apply the `funnyProd()` function to the `(2,4)` arguments first and then to the `(3,5)` ones and then collect the two results in the array `[8,15]`). The problem is that this would not be valid Python syntax.

In cases like this, when you can't simply call a Julia function using Python syntax, you can still rely to the same Julia `eval` function you used to *define* the Python function to also *call* it: `jl.eval("funnyProd.([2,3],[4,5])")`.

Finally, you can access any module available in Julia with `from julia import ModuleName`, and in particular you can set and access global Julia variables using the `Main` module.

7.4 Julia ⇌ R

As you did with Python, with R, you can either run R code (and access its huge set of specialized data science libraries) in Julia (with RCall (`https://github.com/JuliaInterop/RCall.jl`)) or, conversely, execute efficient Julia code from within R (with JuliaCall—see `https://github.com/Non-Contradiction/JuliaCall`).

When you add the `RCall.jl` package, it should automatically download and install R if it doesn't detect a local R installation (with R >= 3.4.0). If this doesn't work for any reason, you can always install R before installing RCall.

You can choose the version of R that `RCall.jl` will use by setting the environmental variable R_HOME to the base directory of your R installation (Use `ENV["R_HOME"]="*"` instead to force `RCall.jl` to download a "private" Julia version of R, even if a local version is available.).

```
ENV["R_HOME"]="/path/to/the/R/base/directory"
# E.g. /usr/lib/R/"                                    JULIA
using Pkg
Pkg.build("RCall")
```

7.4.1 Interactive R Prompt

You can access an embedded R prompt from Julia with the dollar sign \$ symbol (and then go back to the Julia REPL with the Backspace key). You can then access (in this embedded R prompt) Julia symbols or expressions using the dollar sign again or using the @rput macro. Conversely, you can retrieve R variables back to Julia with the @rget macro:

```
julia> a = [1,2]; b = 3                                    JULIA
julia> @rput a      # Explicitly transfer the `a` variable and
                      the object to which it is bonded to R
R>      c = a + $b   # Access a and, through the dollar symbol,
                      b, in R
julia> @rget c      # Explicitly transfer the `c` variable and
                      the object to which it is bonded to Julia
julia> c            # [4,5]
```

You don't need to convert the data, as RCall supports automatic conversions to and from the basic Julia types (including arrays and dictionaries), as well as popular tabular data formats, such as as DataFrames.

7.4.2 Embed R Code in a Julia Program

Embedding R code in a Julia program is similar to what you saw with Python. You both define and call the R functions with the R"..." string macro, and in the function call, you can directly use your Julia data:

```
using RCall                                                JULIA

R"""
sumMyArgs <- function(i,j) i+j
getNElement <- function(n) {
```

```
  a <- c(0,1,2,3,4,5,6,7,8,9)
  return(a[n])
}
"""
i = [3,4]
a = rcopy(R"sumMyArgs"(3,4))        # 7
b = rcopy(R"sumMyArgs"(i,[5,6]))    # [8,10]
b = rcopy(R"sumMyArgs($i,c(5,6))")  # [8.0,10.0] (alternative)
c = rcopy(R"sumMyArgs"([3,4],5))    # [8,9]
d = rcopy(R"getNElement"(1))        # 0
```

As a notable difference with PyCall, the result of calling the R functions defined with R"..." are not yet converted to exploitable Julia objects, but remain as RObjects. In order to convert the RObjects to standard Julia objects, you can use the rcopy function and, if you want to force a specific conversion, you can still rely on convert:

```
convert(Array{Float64}, R"sumMyArgs"(i,[5,6])) # [8.0,9.0] JULIA
convert(Array{Int64}, R"sumMyArgs($i,c(5,6))") # [8,9]
```

Finally, with R, you don't have to worry about the indexing convention, as R and Julia both adopt one-base array indexing.

7.4.3 Use R Libraries

Although RCall provides two specialized macros for working with R libraries—@rlibrary and @rimport—these are inefficient, as data would be copied multiple times between R and Julia.

Just use the R"..." string macro instead:

```
mydf = DataFrame(deposit = ["London","Paris","New-York",
"Hong-Kong"]; q = [3,2,5,3]) # Create a DataFrame ( a
tabular structure with named cols) in Julia        JULIA
R"""
```

```
install.packages('ggplot2', repos='http://cran.us.r-project.org')
library(ggplot2)
ggplot($mydf, aes(x = q)) +
geom_histogram(binwidth=1)
"""
```

Note   In MacOS, you may need the *Xquartz Windows Server*
(see `https://www.xquartz.org`) to display the chart.

7.4.4 JuliaCall: Using Julia in R

To go the other way around, and embed Julia code within an R workflow,
you can use the R package called JuliaCall (see `https://github.com/Non-`
`Contradiction/JuliaCall`).

Installation

```
> install.packages("JuliaCall)                                    R
```

That's all.

Usage

```
> library(JuliaCall)                                              R
> julia_setup()
```

Note that, unlike with PyJulia, the "setup" function needs to be
called every time you start a new R section, not just when you install the
JuliaCall package. If you don't have `julia` in the path of your system, or
if you have multiple versions and you want to specify the one to work with,
you can pass the `JULIA_HOME = "/path/to/julia/binary/executable/`
`directory"` parameter (e.g., `JULIA_HOME = "/home/myUser/lib/`
`julia-1.1.0/bin"`) to the `julia_setup` call.

JuliaCall depends on some things (like object conversion between Julia and R) from the Julia RCall package. If you don't already have it installed in Julia, it will try to install it automatically.

As expected, JuliaCall also offers multiple ways to access Julia in R.

Let's assume you have all your Julia functions in a file. You are going to reuse the juliaScript.jl script you used in PyJulia:

```julia
function helloWorld()                                          JULIA
    println("Hello World!")
end
function sumMyArgs(i, j)
  return i+j
end
function getNElement(n)
  a = [0,1,2,3,4,5,6,7,8,9]
  return a[n]
end
```

You can access its functions in R with:

```R
> julia_source("juliaScript.jl") # Include the file            R
> julia_eval("helloWorld()") # Prints `Hello World!` and
  returns NULL
> a <- julia_call("sumMyArgs",c(1,2,3),c(4,5,6)) # Returns
  `[1] 5 7 9`
> as.integer(1) %>J% getNElement -> b # Returns `0`, the
  "first" element for both Julia and R
```

Concerning the last example, it highlights the pipe operator, which is very common in R (you will see an equivalent operator in Julia in Chapter 11, "Utilities"). The %>J% syntax is a special version of it, provided by JuliaCall, that allows you to mix Julia functions in a left-to-right data transformation workflow.

You can otherwise embed Julia code directly in R using the julia_
eval() function:

```R
> funnyProdR <- julia_eval('
+    function funnyProd(is, js)
+      prod = 0
+        for i in 1:is
+          for j in 1:js
+            prod += 1
+          end
+        end
+        return prod
+      end
+ ')
```

You can then call this function in R, either as funnyProdR(2,3),
julia_eval("funnyProd(2,3)") or julia_call("funnyProd",2,3).

While other "convenience" functions are provided by the package,
using julia_eval and julia_call should be sufficient to accomplish any
task you need in Julia.

CHAPTER 8

Effectively Write Efficient Code

This chapter includes a discussion of the following third-party packages:

Package Name	URL	Version
BenchmarkTools.jl	https://github.com/JuliaCI/BenchmarkTools.jl	v0.4.2
ProfileView.jl	https://github.com/timholy/ProfileView.jl	v0.4.1
JuliaInterpreter.jl	https://github.com/JuliaDebug/JuliaInterpreter.jl	v0.6.1
Juno	https://junolab.org	
	https://github.com/JunoLab/Atom.jl	v0.9.1
	https://github.com/JunoLab/atom-julia-client	v0.9.4

The objective of this chapter is to present topics that, while maybe not strictly essential, are nevertheless very important for writing efficient code in an efficient manner, i.e. code that runs fast and is easy (and fun) to write. I first deal with code performance, then move to parallelization in order to speed up execution even more. I continue discussing tools that allow developers to check that the program is following the intended behavior

A. Lobianco, *Julia Quick Syntax Reference*, https://doi.org/10.1007/978-1-4842-5190-4_8

and facilitate early bug determination. I finish the chapter with the constructs that you can employ to isolate, and possibly correct, problems that arise at runtime.

8.1 Performance

Julia is already relatively fast when working with data objects of type Any, i.e. when the type of the object remains unspecified until execution. But when the compiler can infer a specific type (or a union of a few types), Julia programs can run with the same order of magnitude as C programs.

The good news is that programmers don't typically need to specify variables types. They are indeed inferred directly by the compiler to produce efficient, specialized code. It is only in rare situations that the compiler can't infer a specific type.

In this section, you see how to measure (or benchmark) Julia code, how to avoid this problem, and see a few other tips, specific to Julia, that can help improve performance.

8.1.1 Benchmarking

When you highlight code in your editor and run it to evaluate the corresponding instructions, Julia performs two separate tasks: it compiles the code (and eventually its dependencies) and then evaluates it.

 Unless you are interested in how effective Julia is for interactive development, you should benchmark and/or profile only the second operation (the evaluation), and not the code compilation.

Hence, if you want to use the simplest benchmarking macro, `@time [a Julia expression]`, you should run it twice and discard the first reading, when compile time is included.

Let's consider for example a function that computes the Fibonacci sequence, where each number is the sum of the Fibonacci function applied to the previous two numbers (i.e., $F_n = F_{n-1} + F_{n-2}$), with F_0 and F_1 defined as 0 and 1, respectively.

A naive recursive implementation would look like this (note that much faster algorithms to compute Fibonacci numbers exist):

```julia
julia> function fib(n)                                    JULIA
           if n == 0 return 0 end
           if n == 1 return 1 end
           return fib(n-1) + fib(n-2)
       end
fib (generic function with 1 method)
```

If you start a new Julia session, you could run this code and try to time it with the @time macro. You would obtain something similar to this:

```julia
julia> @time fib(10)                                      JULIA
  0.004276 seconds (4.63 k allocations: 254.928 KiB)
55
julia> @time fib(10)
  0.000006 seconds (4 allocations: 160 bytes)
55
```

While the @time macro can accept any valid Julia expression, it's most often used with function calls.

You can see that the first time and memory usage information reported includes the compilation of the function (You can also deduce that the compilation of a function happens "just in time" (JIT), which is the first time the function is called, and not when it is defined.).

Further, is the function time *really* 0.000006 seconds? That's only one example of a variable that includes some stochastic components related to the status of the machine. A better approach is to have some *average* timing, with a number of draws that makes the mean statistically significant.

The BenchmarkTools package (see https://github.com/JuliaCI/ BenchmarkTools.jl) does exactly that: it provides @benchmark, an alternative macro that automatically skips the compilation time and runs the code several times in order to report an accurate timing.

Applied to this function, you would have the following:

```julia
julia> @benchmark fib(10)                              JULIA
BenchmarkTools.Trial:
  memory estimate:  0 bytes
  allocs estimate:  0
  --------------
  minimum time:     292.131 ns (0.00% GC)
  median time:      292.401 ns (0.00% GC)
  mean time:        299.772 ns (0.00% GC)
  maximum time:     492.895 ns (0.00% GC)
  --------------
  samples:          10000
  evals/sample:     267
```

This gives you a much more precise benchmark result.

8.1.2 Profiling

While benchmarking gives information about the *total* time spent running the given instructions, often you want to know how the time is allocated within the function calls, in order to find bottlenecks.

This is the task of a *profiler*. Julia comes with an integrated statistical profiler (in the Profile package of the standard library). When invoked with an expression, the profiler runs the expression and records, every x milliseconds, the line of code where the program is at that exact moment. The more often a line of code is hit, the more computationally expensive it is.

Using this sampling method, at the cost of losing some precision, profiling can be very efficient. It will lead to a very small overhead compared to running the code without it. And it is very easy to use:

- `@profile myfunct()`: Profile a function (remember to run the function once before, or you will also measure the JIT compilation).

- `Profile.print()`: Print the profiling results.

- `Profile.clear()`: Clear the profile data.

Profile data is accumulated until you clear it. This allows developers to profile even very small (fast) functions, where a single profiling run would be inaccurate. Just profile them multiple times, such as:

```
@profile (for i = 1:100; foo(); end)
```

For example, let's assume that you have the following function to profile in a file called myCode.jl, with the numbers on the left being the line numbers:

```
64 function lAlgb(n)                                          JULIA
65    a = rand(n,n) # matrix initialization with random numbers
66    b = a + a      # matrix sum
67    c = b * b      # matrix multiplication
68 end
```

If you run `@profile (for i = 1:100; lAlgb(1000); end)` (again, after having run the function once for the JIT compilation) and then run `Profile.print()`, you may end up with something like the following output (that has been partially cut for space reasons):

```
julia> Profile.print()                                        JULIA
1627 ./task.jl:259; (::getfield(REPL, Symbol("##26#27"))
{REPL.REPLBackend})()
1627 ...ld/usr/share/julia/stdlib/v1.1/REPL/src/REPL.jl:117;
macro expansion
```

```
1627 ...obianco/.julia/packages/Atom/E4PBh/src/repl.jl:135;
repleval(::Module, ::Expr)
 1627 ./boot.jl:328; eval
   1627 ...share/julia/stdlib/v1.1/Profile/src/Profile.jl:25;
   top-level scope
    1627 ./none:1; macro expansion
     267 /full/path/to/myCode.jl:65; lAlgb(::Int64)
       267 .../share/julia/stdlib/v1.1/Random/src/Random.jl:243;
       rand
     443 /full/path/to/myCode.jl:66; lAlgb(::Int64)
      443 ./arraymath.jl:47; +(::Array{Float64,2},
      ::Array{Float64,2})
       440 ./broadcast.jl:707; broadcast(::typeof(+),
       ::Array{Float64,2}, ::Array{Float64,2})
       1 ./int.jl:0; broadcast(::typeof(+), ::Array{Float64,2},
       ::Array{Float64,2})
       2 ./simdloop.jl:71; broadcast(::typeof(+),
       ::Array{Float64,2}, ::Array{Float64,2})
     917 /full/path/to/myCode.jl:67; lAlgb(::Int64)
      181 ./boot.jl:404; *
      736 ...julia/stdlib/v1.1/LinearAlgebra/src/matmul.jl:142; *
```

The format of each line in the output is the number of samples in the corresponding line and in all downstream code, the filename, the line number, and then the function name.

The important lines are those related to the individual function's operation. From the profile log, you see that the initialization of the matrix with a random number (row 65) has been hit 267 times, the matrix summation (row 66) 443 times, and the matrix multiplication (row 67) 917 times.

This helps you determine the *relative* time spent within the lAlgb function, with the matrix multiplication being the most computationally intensive operation.

In real use, things quickly become complicated, and you may appreciate a graphical representation of the call stack with the profiled data. You may then want to use the `ProfileView` package. With `ProfileView.view()`, it provides a nice chart of the call stack, with the width proportional to the hit count.[1]

If you use a recent version of Juno as your Julia IDE, you have some convenient user interfaces as well, as shown in Figure 8-1 (you still need to use the `@profile` macro to collect the data you are interested in):

- `Juno.profiletree()` provides a nice replacement for `Profile.print()`, where the tree items are collapsible/expandable on click and linked to the relative source code.

- In a similar way, `Juno.profiler()` improves over `ProfileView.view()` by showing the call chart in the panel and removing noise due to using Atom, focusing on the call stack of the user code.

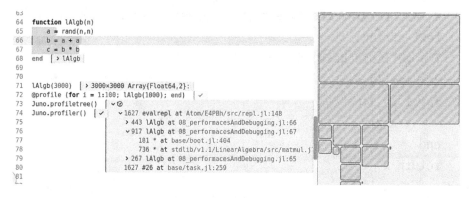

Figure 8-1. *Profile tools in the Atom IDE*

---

[1]At the time of the writing, `ProfileView` seems to have trouble installing on MacOS.

8.1.3 Type Stability

A function is said to be type-stable when its return type depends only on the types of the input, not on their values. While type-unstable functions have their return type compiled as Any, and only at runtime are they converted to a specific type, type-stable functions can be optimized by the compiler, as it is able to compile specialized versions for all the possible types the function can be called with. Further, type stability is essential in retaining the capacity of the compiler to infer the types of the various objects across the various chained calls.

It is important to note that type stability is *not* the problem that parameters aren't annotated with a specific type. Often the compiler will still be able to list, at compile time, all the possible types of the parameters for that function, and still produce performant code, even without explicit type annotation.

Let's look at an example of a type-unstable function:

```julia
function fUnstable(s::AbstractString, n)
    # This returns the type that is given in parameter `s` as a
    text string:
    T = getfield(Base, Symbol(s))
    x = one(T)
    for i in 1:n
        x = x+x
    end
    return x
end
```

The problem here is that the returned value depends on the specific text entered as input, that is, on the *value* of the s parameter. While this may seem like an extreme case, type-unstable functions often arise when processing user data or reading from some inputs.

Type stability of a function can be checked with the @code_warntype
myfunct() macro.

If you run @code_warntype fUnstable("Float64",100000) in a
terminal, you would obtain a lengthy output, but the important bit is the
first line that would read as Body::Any (displayed in red in the Julia REPL).
It is this information that tells you that the function is type-unstable.

One way to remove type instability in this example is to fix the x
variable to Float64:

```julia
function fStable(s::AbstractString, n)
    T = getfield(Base, Symbol(s))
    x = one(Float64)
    for i in 1:n
        x = x+x
    end
    return x
end
```

If you run @code_warntype fStable("Float64",100000) you receive
in the first line of the output a Body::Float64 (displayed in a reassuring
blue), which confirms the type stability of the function.

You can benchmark the two functions. Improvements due to type
stability are huge:

```julia
@benchmark fUnstable("Float64",100000) # median time: 1.299 ms
@benchmark fStable("Float64",100000) # median time: 111.461 μs
```

Finally, if there are functions that really cannot be made type-stable,
a strategy could be to break the function into multiple definitions, where
hopefully the computational-intensive part can be made type-stable.

8.1.4 Other Tips to Improve Performance

Avoid Using Global Variables and Run Performance-Critical Code Within Functions

Global variables (variables that are defined outside any function) can change their type at any time.

Instead of using a global variable inside a function, pass the variable as a function parameter. If this is not possible or practical, try at least to declare the global variable as const (as to fix its type) or annotate its type at the point of usage in the function.

Annotate the Type of Data Structures

As an exception to the rule that annotation in Julia is generally useless for performance, to annotate the type of the inner elements of a data container with a concrete type can help with performance.

Given that primitive types have fixed memory size, an array of them (e.g., Array{Float64}) can be stored in memory contiguously, in a highly efficient way. Conversely, containers of Any elements, or even any abstract type, are stored as pointers to the actual memory that ends up being sparsely stored and hence relatively slow to be accessed.

Annotate the Fields of Composite Types

As a further exception to the annotation-is-not-needed rule, leaving the field of a structure unannotated (or annotated with an abstract type) will leave type-unstable all the functions that use that structure. If you want to keep the flexibility, use parametric structures instead. For example:

JULIA

```julia
struct MyType2
  a::Float64
end
```

```
# ..or..
struct MyType3{T}
  a::T
end
# ..rather than..
struct MyType
  a
end
```

Loop Matrix Elements by Column and Then by Row

Matrix data is stored in memory as a concatenation of the various column arrays (*column-major* order), i.e. the first column, followed by the second column, etc.

It is hence faster to loop matrices first over their columns and then over their rows (as to keep reading contiguous values):

```
M = rand(2000,2000)                              JULIA
function sumRowCol(M) # slower
    s = 0.0
    for r in 1:size(M)[1]
        for c in 1:size(M)[2]
            s += M[r,c]
        end
    end
    return s
end
function sumColRow(M) # faster
    s = 0.0
    for c in 1:size(M)[2]
        for r in 1:size(M)[1]
            s += M[r,c]
```

```
        end
    end
    return s
end
@benchmark sumRowCol(M) # median time:    21.684 ms
@benchmark sumColRow(M) # median time:    4.787 ms
```

8.2 Code Parallelization

Julia provides core functionality to facilitate code parallelization, either using multiple threads or multiple processes. Threads share the same memory (so there is no need to copy data across multiple threads), but are constrained to run on the same CPU. Also, it is difficult to "guarantee" safe multi-threading. Thread support in Julia is still experimental, although its development is advancing fast[2].

Processes work each on their own memory (hence, the data needs to be copied across processes, representing a computational cost to balance with the advantages given by the parallel execution) and can spawn on multiple CPUs or even on multiple machines (e.g., through SSH connections). Because multi-process parallelization in Julia is mature and stable, in the rest of this section, I will discuss it. All the functions you see are part of the Distributed standard library package.

8.2.1 Adding and Removing Processes

Processes can be added and removed at any time:

```
wksIDs = addprocs(3) # 2,3,4                                    JULIA

println("Worker pids: ")
for pid in workers()
```

---

[2]Comprehensive thread API will be introduced with Julia 1.3.

```
    println(pid) # 2,3,4
end

rmprocs(wksIDs[2]) # or rmprocs(workers()[2]) remove process pid 3

println("Worker pids: ")
for pid in workers()
    println(pid) # 2,4 left
end

@everywhere println(myid()) # 2,4
```

The first row adds three processes. In this case, it adds them on the same machine as the one running the main Julia process, but it is possible to specify in addprocs() the SSH connection details directly in order to add the processes to other machines (Julia must be installed on those machines as well).

Generally it makes sense for n to equal the number of cores available on the machine.

It is also possible to start Julia directly with multiple processes using the command-line flag -p (e.g., ./julia -p 2). The -p argument implicitly loads the Distributed module.

In this example, the "internal IDs" of the new processes (different from the OS-level assigned pids) are saved in the array wksIDs, although you don't really need it, as a call to workers() would return that array as well.

The main Julia process (e.g., the one providing the interactive prompt) always has a pid of 1 and, unless it is the sole process, it is not considered a worker.

You can then check the working processes that are active and remove those that you no longer need with rmprocs(pid).

Finally, you can use the function myid() to return the PID of a process and spread the function to print it across all the working processes using the @everywhere macro.

8.2.2 Running Heavy Computations on a List of Items

This section considers two common patterns of parallelization.

The first one involves running a computationally expensive operation over a list of items, where you are interested in the result of each of them. Critically, the computation of each item is independent of the other items.

As an example, you are going to reuse the fib function you saw earlier. You first need to be sure that all processes (assume you added three working processes with addprocs(3)) know about the function you want them to use. You can use the @everywhere macro in front of the definition of the function in order to transfer the function to all the workers (if you add more processes at a later stage, they will not have access to such a function). If you have multiple functions to invoke in the working threads (or there are several sub-functions to call), a convenient option is to define all of them and the eventual global variables they need in a begin block prefixed with the @everywhere macro (i.e., @everywhere begin [shared functions definitions] end) or put them in a file and run @everywhere include("computationalcode.jl").

You can then create the array of input data and run the heavy operation on each element of it with pmap(op,inputData):

```julia
a = rand(1:35,100)
@everywhere function fib(n)
    if n == 0 return 0 end
    if n == 1 return 1 end
    return fib(n-1) + fib(n-2)
end
@benchmark results = map(fib,a)  # serialized: median time:
490.473 ms
@benchmark results = pmap(fib,a) # parallelized: median time:
249.295 ms
```

The pmap function automatically and dynamically picks up the "free" processes, assigns them the job, and merges the results in the returned array. Only the evaluation of the function is done in the workers.

pmap is convenient when the computation cost of the function is high, like in this case. Note, however, that even if you are using three working threads, computational time is divided by a factor well below three[3].

8.2.3 Aggregate Results

The second situation is when you want to perform a small operation on each of the items but you also want to perform an "aggregation function" at the end to retrieve a scalar value (or an array if the input is a matrix).

In these cases, you can use the @distributed (aggregationFunction) for construct. As an example, you run in parallel a division by 2 and then use the sum as the aggregation function (assume three working processes are available):

```julia
function f(n)                                          JULIA
  s = 0.0
  for i = 1:n
    s += i/2
  end
    return s
end
function pf(n)
  s = @distributed (+) for i = 1:n # aggregate using sum on
                                  variable s
```

---

[3]This is known in computer science as *Amdahl's Law*, and it is due to bottlenecks from the serialized part of the code and the computational costs of the parallelization.

```
        i/2                             # last element of for cycle
                                        is used by the aggregator
   end
   return s
end
@benchmark  f(10000000) # median time:    11.478 ms
@benchmark pf(10000000) # median time:     4.458 ms
```

The two cases exposed are some of the most frequently encountered in code parallelization. Julia offers a fairly complete list of high- and low-level functions for parallelization, and some external packages, like `DistributedArrays` (see `https://github.com/JuliaParallel/DistributedArrays.jl`), help you achieve efficient parallelized code for many situations in a relatively simple way.

It is worth noticing that parallel computing in Julia can be extended to work on the GPU (Graphical Processing Unit) rather than the CPU. The topic goes beyond the scope of this book, but you can check the official documentation (see `https://docs.julialang.org/en/v1/manual/parallel-computing/index.html`), where several packages, providing both low-level and high-level interfaces, are proposed.

8.3 Debugging

Debugging is, in general, "the process of eliminating errors or malfunctions in a computer program" (from the Merriam-Webster dictionary).

Debugging is much harder in compiled languages, and this prompted the development of many sophisticated, dedicated tools to help debugging programs.

The interactive and introspective nature of Julia makes debugging relatively simple, as the programmer can easily analyze, and eventually change, the code at any step. Still, specialized debugging tools have been

developed to help debug code within functions that would otherwise be
difficult to analyze in their individual components.

8.3.1 Introspection Tools

Introspection is the ability of a program to examine itself, and in particular,
type introspection is the ability of a program to examine the type or
properties of an object at runtime.

The first "introspection" tool is, in a wide sense, the Workspace panel,
which is available in several Julia development environments (Figure 8-2
shows the one available in Juno). It shows all the identifiers (variable,
functions, and so on) available in the current scope and the objects bound
to them.

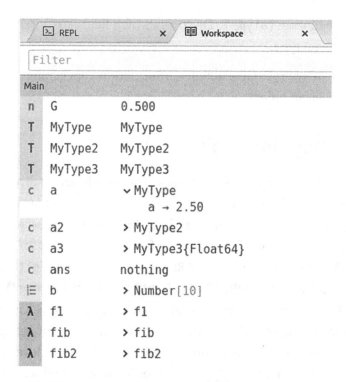

Figure 8-2. *Workspace panel in the Juno IDE*

In addition, Julia offers the following introspection functions:

- `methods(myfunction)`: Retrieves all the function signatures, that is, all the different types of parameters the function accepts when it is called.

- `@which myfunction(myargs)`: Discovers which method would be used in a specific call (within the several available, as Julia supports multiple-dispatch).

- `typeof(a)`: Discovers the type of a given object.

- `eltype(a)`: Discovers the type of the inner elements of a given object, e.g. `typeof(ones(Int64,10))` is `Array{Int64,1}` and `eltype(ones(Int64,10))` is `Int64`.

- `fieldnames(aType)`: Discovers which fields are part of a given type.

- `dump(myobj)`: Gets detailed information concerning the given object.

8.3.2 Debugging Tools

Julia offers a full-featured debugger stack. The base functionality is provided by the `JuliaInterpreter.jl` package (see `https://github.com/JuliaDebug/JuliaInterpreter.jl`), while the user interface is provided by the command-line packages `Debugger.jl` (`https://github.com/JuliaDebug/Debugger.jl`) and `Rebugger.jl` (`https://github.com/timholy/Rebugger.jl`), or the Juno IDE (`https://junolab.org/`).

Since these functionalities are similar, and perhaps using a graphical user interface is a bit more comfortable, let's discuss only the integration of the debugger in Juno. The Juno integrated debugger allows developers to inspect the source code at any point and set *breakpoints*, which are points where the execution halts so the programmer can inspect the situation, change the code, or continue.

Let's reuse the fib function and open a debugging session for it (see Figure 8-3).

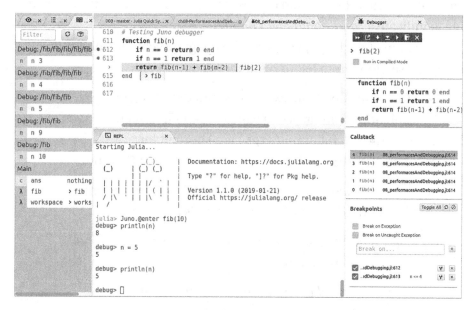

Figure 8-3. *Debugging session in the Juno IDE*

You can start a debugging session by typing Juno.@enter function(args) in the Juno REPL. In this case, you can use Juno.@enter fib(10).

If it isn't already visible, starting the debugging session will open the specific Debugger panel (shown on the right side in Figure 8-3). The REPL prompt also changes to debug>.

From Figure 8-3, you can see that you used the debug prompt to print the value of the variable n (but you already knew it from the Workspace panel on the left) and, most interesting, to dynamically change its value. You used it to run an odd Fibonacci sequence where you skipped the recursions from 8 to 6.

Clicking on the left side of the line numbers in the main Editor panel will toggle a breakpoint on that line (a red dot). Breakpoints are listed on the bottom of the Debugger panel. They can be disabled (so they are no longer active) or can be edited so that they apply only when certain conditions are met (here, when n is less or equal to 4). The dot will become blue to indicate a *conditional breakpoint*.

On the top of the Debugger panel are the graphical controls for the debugging session. From left to right, these controls are:

- *Continue:* Continue execution until the debugged function terminates or a valid breakpoint is met. `Juno.@run function(args)` is a shortcut for `Juno.@enter function(args)` plus pressing Continue.

- *Finish Function:* Continue execution until the current function exits or a valid breakpoint is met.

- *Next Line:* Evaluate the current line and stop just before evaluating the following line.

- *Step to Selected Line:* Continue execution until the selected line or a valid breakpoint is met.

- *Next Expression:* Evaluate the following expression. Differs from Next Line only if multiple expressions are defined in the same line.

- *Step Into Function:* Execute the next function call and stop at the beginning of the sub-function (e.g., in Figure 8-3, it would call `fib(2)` and halt at its start). The function that would be called is shown just under the controls.

- *Stop Debugging:* Terminate the debugging session and pass the control to the normal REPL.

Finally, note the Callstack session in the middle of the Debugger panel: it provides a clickable history of the chain of calls that led to the current state.

8.4 Managing Runtime Errors (Exceptions)

Runtime errors can be handled, as in many other languages, with a try/catch block:

```julia
try
  # ..some dangerous code..
catch
  # ..what to do if an error happens, for example send an error
  message using:
  error("My detailed message")
end
```

You can also check for a *specific* type of exception. In the following snippet, you use exceptions to return missing if a specific key is not found in a dictionary representing some data:

```julia
function volume(region, year)
    try
        return data["volume",region,year]
    catch e
        if isa(e, KeyError)
          return missing
        end
        rethrow(e)
    end
end
```

PART II

Packages Ecosystem

CHAPTER 9

Working with Data

This chapter includes a discussion of the following third-party packages:

DataFrames.jl	https://github.com/JuliaData/DataFrames.jl	v0.19.2
DataFramesMeta.jl	https://github.com/JuliaStats/DataFramesMeta.jl	v0.5.0
Query.jl	https://github.com/JuliaDebug/JuliaInterpreter.jl	v0.12.1
IndexedTables.jl	https://github.com/JuliaComputing/IndexedTables.jl	v0.12.2
LAJuliaUtils.jl	https://github.com/sylvaticus/LAJuliaUtils.jl	v0.2.0
Pipe.jl	https://github.com/oxinabox/Pipe.jl	v1.1.0
Plots.jl	https://github.com/JuliaPlots/Plots.jl	v0.26.1
StatsPlots.jl	https://github.com/JuliaPlots/StatsPlots.jl	v0.10.2

Although Julia already natively provides a fast structure for working with tabular (and multidimensional) data, the `Array{T,n}` parameterized type, a very large collection of third-party packages has been created to work specifically with numerical data.

This happened, in general terms, because of two needs.

© Antonello Lobianco 2019
A. Lobianco, *Julia Quick Syntax Reference*, https://doi.org/10.1007/978-1-4842-5190-4_9

The first one was better convenience. Packages like DataFrames or IndexedTables add metadata like column names, allowing developers to retrieve data by name instead of by position, query it with `Query.jl` or `DataFramesMeta.ij`, and then join, group, chain operations with `Pipe.jl`, etc.

The second need was for efficient heterogeneous data. In a Julia base `Array{T,2}` all data of the matrix must be of the same type, for efficiency's sake. The alternative is to work with `Array{Any,2}` and accept lower performance.

With the aforementioned packages, different columns can have different types, a situation rather common when working with datasets, and the computations involving these data structures remain performant.

This chapter presents the main Julia third-party packages for data manipulation, including data structures, various query and manipulation tools, and plotting and visualization tools.

9.1 Using the DataFrames Package

DataFrames are the first data structure you see, and perhaps the most used one. They are very similar to R's dataframes and Python's Pandas. The approach and the function names are very similar too, although the way of accessing the API may be a bit different. Unlike Pandas, the DataFrames package works only with two-dimensional (tabular) data.

Internally, a DataFrame is a collection of standard arrays, each of its own type `T`, or eventually of type `Union{T,Missing}` if missing data is present.

9.1.1 Installing and Importing the Library

You can install the library in the usual way:

- Install the library: `] add DataFrames`
- Load the library: `using DataFrames`

Some functions presented in this section use LAJuliaUtils (see https://github.com/sylvaticus/LAJuliaUtils.jl), my personal repository of utility functions.

As this is not a registered Julia package, install it with:

```
] add https://github.com/sylvaticus/LAJuliaUtils.jl.git
```

9.1.2 Creating a DataFrame or Loading Data

These are various ways to create or load a DataFrame ("df" for short). Methods for loading a df from a comma-separated or Excel file were discussed in Chapter 5, in the section "Parsing CSV Files," as DataFrame is the default output format.

To create a df from scratch:

ℹ Note the following (fictional) database of timber markets: I will refer to it across the whole chapter.

```
df = DataFrame(region      = ["US","US","US","US","EU",
                              "EU","EU","EU"],                 JULIA
               product     = ["Hardwood","Hardwood","Softwood",
                              "Softwood","Hardwood","Hardwood",
                              "Softwood","Softwood"],
               year        = [2010,2011,2010,2011,2010,2011,
                              2010,2011],
               production  = [3.3,3.2,2.3,2.1,2.7,2.8,1.5,1.3],
               consumption = [4.3,7.4,2.5,9.8,3.2,4.3,6.5,3.0])
```

To create an empty df:

```
df = DataFrame(A = Int64[], B = Float64[])
```

To convert a df from a matrix of data and a (separate) vector of column names:

```
mat = [1 2 3; 4 5 6]
headerstrs = ["c1", "c2", "c3"]
df = DataFrame([[mat[:,i]...] for i in 1:size(mat,2)], Symbol.
(headerstrs))
```

To convert a df from a matrix, where the headers are defined in the first row of the matrix:

```
mat = ["c1" "c2" "c3"; 1 2 3; 4 5 6]
df = DataFrame([[mat[2:end,i]...] for i in 1:size(mat,2)],
Symbol.(mat[1,:]))
```

To create a df from a table defined in code:

```
using CSV                                                        JULIA
df = CSV.read(IOBuffer("""
region product  year production consumption
US      Hardwood 2010 3.3          4.3
US      Hardwood 2011 3.2          7.4
US      Softwood 2010 2.3          2.5
US      Softwood 2011 2.1          9.8
EU      Hardwood 2010 2.7          3.2
EU      Hardwood 2011 2.8          4.3
EU      Softwood 2010 1.5          6.5
EU      Softwood 2011 1.3          3.0
"""),delim=" ", ignorerepeated=true, copycols=true)
```

ⓘ Without the `copycols=true` parameter, the resulting DataFrame would be available in read-only mode.

To import a df from an OpenDocument spreadsheet file (a format used by OpenOffice, LibreOffice, MS Excel, and others), use the ods_read() function from the OdsIO package (https://github.com/sylvaticus/OdsIO.jl), together with the retType="DataFrame" argument and the top-left/bottom-right range:

```julia
df = ods_read("spreadsheet.ods";sheetName="Sheet2",retType=
"DataFrame",range=((tl_row,tl_col),(br_row,br_col)))      JULIA
```

9.1.3 Getting Insights About the Data

Once you have a DataFrame, you may want to analyze its data or its structure:

- show(df): Show an extract of the df (depending on the space available). Use allrows=true and allcols=true to retrieve the whole df, whatever its size.

- first(df,n): Show the first *n* lines of the df.

- last(df, 6): Show the last *n* lines of the df.

- describe(df): Retrieve type and basic descriptive statistics about each column of the df.

- names(df): Return an array of column names (as symbols, not as strings).

- for r in eachrow(df): Iterate over each row.

- for c in eachcol(df): Iterate over each column.

- unique(df.fieldName) or [unique(c) for c in eachcol(df)]: Return the unique elements of the columns.

- [eltype(c) for c in eachcol(df)]: Return an array of column types.

- size(df) (rows,cols), size(df)[1] (rows), size(df) [2] (cols): Return the size of the df.

Note that column names are Julia symbols, not strings. To programmatically compose a column name, you need to use the `Symbol(aString)` constructor. For example, `Symbol("value_",0)`.

9.1.4 Filtering Data (Selecting or Querying Data)

To select (filter) data, you can use the methods provided by the DataFrame package or specialized methods provided by other packages, such as `DataFrameMeta.jl` or `Query.jl`.

In particular, the `Query.jl` package aims to provide a generic API that is valid regardless of the data backend.

- To select whole columns:
  - `df[:,:cNames]`: Select multiple columns (e.g., `df[:,[:product,:year] ]`) by copying the data.
  - `df[!,:cNames]`: Select multiple columns (e.g., `df[!,[:product,:year] ]`) by referencing the data.
  - `df.cName`: Select an individual column (e.g., `df.product`) (referenced, i.e., equal to `df[!,:cName]`).
  - `df[cPos]`: Select one or more columns (e.g., `df[2]`) by copying the data.
- Select whole rows:
  - `df[rPos,:]`: Select one or more rows (e.g., `df[2,:]`). Unlike columns, rows have no title name.
- Select cells:
  - `df.cName[rPos]`: Select specific cells in the `cName` column (e.g., `df.product.[1:2:6]`).
  - `df[rPos, :cName(s)]` or `df[rPos,cPos]`: Select specific cells by row position and column names or position (e.g., `df[2,:product]`).

Where:

- cName are individual column names (symbols).

- cNames are an array of column names (symbols).

- rPos are the row positions (scalar, array, or range).

- cPos are the column positions (scalar, array, or range).

The returned selection is:

- An Array{T,1} if it is a single column

- A DataFrameRow (similar to a DataFrame) if it's a single row

- T if it's a single cell

- Another DataFrame otherwise

Filtering:

- df[arrayOfBooleans,:]: Select the rows where arrayOfBooleans (that must of congruent size) is true. This is known as a "Boolean selection".

  This also works with column positions to select rows and columns at the same time (but not for a matrix of Boolean values to use as a mask). Some examples:

  - df[[true,true,false,false,false,false, false,false],[true,true,true,false,true]] selects the first two rows of all the columns except the fourth one.

  - df[[i in ["US","China"] for i in df.region], :] filters by value, based on a value being in a list using Boolean selection through list comprehension. The list comprehension part

returns an array of Boolean values based on the
condition, then you use the Boolean selection over it.

- `df[([i in [“US”,”China”] for i in df.region]`
  `.> 0) .& (df.year .== 2010), :]` filters by
  multiple conditions in the Boolean selection. The
  dot is needed to vectorize the operation. Note the
  use of the bitwise and the single ampersand.

- `df[startswith.(df.region,”E”),:]` filters based
  on the initial pattern of the desired values.

- `@where(df, condition(s))`: Filter using `@where` (from
  the DataFrameMeta package). In case of multiple
  conditions, all of them must be verified. If the column
  name is stored in a variable, you need to wrap it
  using the `cols()` function. For example, `colToFilter`
  `= :region; @where(df, :production .> 3,`
  `cols(colToFilter) .== "US")`.

Another (and perhaps more elegant, although longer) way to query
a DataFrame is to use the Query package (see `https://github.com/`
`queryverse/Query.jl`). The first example selects a subset of rows and
columns; the second one highlights how you can mix multiple selection
criteria:

```julia
dfOut = @from i in df begin                                    JULIA
          @where i.region == "US"
          # Select a group of columns, eventually changing
          their name:
          @select {i.product, i.year, USProduction=i.production}
          @collect DataFrame
        end
dfOut = @from i in df begin
          @where i.production >= 3 && i.region in ["US","China"]
```

```
    @select i # Select the whole rows
    @collect DataFrame
end
```

9.1.5 Editing Data

There are many ways in which you can edit data:

- df[rowIdx,:col1] .= aValue: Edit values
  by specifying rows and columns (e.g.,
  df[[1,2],:production] .= 4.2).

- df[(df.col1 .== "foo"), :col2] .= aValue:
  Change values by filtering through a Boolean selection
  (e.g., df[(df.region .== "US") .& (df.product .==
  "Hardwood"), :production] .= 5.2).

- df.col1 = map(akey->myDict[akey], df.col1):
  Replace values (or add columns) based on a dictionary
  (the original data—df.col1—could be in a different
  column or in a different DataFrame, e.g. rFullName =
  Dict("US" => "United States", "EU" => "European
  Union"); df.regFullName = map(rShortName-
  >rFullName[rShortName], df.region)).

- df.newCol = df.col1 .* " " .* df.col2:
  Concatenate (string) values for several columns
  to create the value in a new column (e.g., df.
  regAndProduct = df.region .* " " .* df.product).

- df.a = df.b .* df.c: Compute the value of an
  existing or new column based on the values of other
  columns. Note that you need to make element-wise
  operations. Sometimes (e.g., with addition and
  subtractions) it is not needed, but in general the

145

operation between the columns should use the dot operator (except for the actual assignation with the equals sign).

Alternatively, you can use map. For example, df.netExport = map((p,c) -> p - c, df.production, df.consumption).

- push!(df, [aRowVector]): Append a row to the df, e.g., push!(df,["EU" "Softwood" 2012 5.2 6.2]).

- deleterows!(df,rowIdx): Delete the given rows. rowIds can be a scalar, a vector, or a range. You can alternatively just copy the df without the rows that are not needed, e.g., i=3; df2 = df[[1:(i-1);(i+1):end],:].

- df = similar(df,0): Empty a df. df2= similar(df,n) copies the structure of a df by adding *n* rows (with rubbish data) to the new DataFrame (*n* defaults to size(df)[1]). You can use it with n=0 in order to empty a given df.

9.1.6 Editing Structure

There are also many ways in which you can edit the structure of the data:

- select!(df,Not([:col1,:col2])): Delete columns by name.

- names!(df, [:newCName1,:newCName2,:newCName3]): Rename all columns. Use rename!(df, Dict(:oldCName1=> :newCName1)) to rename only a few columns.

- df = df[:,[:b, :a]]: Change the column order.

- `df.id = 1:size(df, 1)`: Add an "id" column (useful for unstacking an operation; we see more of this later in this chapter).

- `addCols!(df, [:colNames], [colTypes])`: Add to the DataFrame empty columns called `colNames` of types `colTypes` (from the LAJuliaUtils package). For example, `addCols!(df, [:import, :export], [Float64, Float64])`.

- `df.a = Array{Union{Missing,Float64},1}(missing,size(df,1))`: Add a `Float64` column (filled with missing values by default).

- `insert!(df, i, [colContent], :colName)`: Insert a column at position i.

- `df.a = convert(Array{T,1},df.a)`: Convert column type to T.

- `df.a = map(string, df.a)`: Convert from Int64 (or Float64) to String (e.g., `df.namedYear = map(string,df.year)`).

- `stringToFloat(str) = try parse(Float64, str) catch; return(missing) end; df.a = map(stringToFloat, df.a)`: Convert from String to Float (converting to integers is similar).

- `categorical!(df, [:a, :b])`: "Pool" specific columns in order to efficiently store repeated categorical variables. Note that while the memory decreases, filtering with categorical values is not necessarily quicker (indeed it can be a bit slower).

- `collect(df.a])`: Convert a categorical array into a normal one.

Merging/Joining/Copying Datasets

You can use these methods to join datasets:

- `df = vcat(df1,df2,df3)`: Vertically concatenate different DataFrames with the same structure (or `df = vcat([df1,df2,df3]...)` (note the three dots at the end, i.e., the splat operator).

- `df = hcat(df1,df2, makeunique=true)`: Horizontally concatenate DataFrames with the same number of rows. `makeunique=true` is needed if the DataFrames to be merged have columns with identical names.

- `fullDf = join(df1, df2, on = :commonCol)`: Join DataFrames horizontally. The `on` parameter can also be an array of common columns. There are many possible types of `join`. The most common ones are `:inner` (the default; only rows with keys on both sides are returned), `:left` (rows on the left df with keys not present on the right df are also returned), `:right` (opposite of `:left`), and `:outer` (rows with elements missing in any of the two dfs are also returned).

- `df2 = similar(df1, 0)`: Copy the structure of a DataFrame (to an empty one).

9.1.7 Managing Missing Values

Recall that missing values (i.e. `missing`, the only instance of the singleton `Missing` type) will propagate silently across operations if they aren't dealt with. The DataFrames package provides a series of methods to allow the programmer to decide what to do with this data, depending on the interpretation of missingness:

- `dropmissing(df)` and `dropmissing!(df)`: Return a df with only complete rows, i.e. without `missing` in any field (`dropmissing(df[[:col1,:col2]])` is also available). Within an operation (e.g., a sum), you can use `dropmissing()` in order to skip `missing` values before the operation takes place.

- `completecases(df)` or `completecases(df[[:col1,:col2]])`: Return a Boolean array concerning absence in the whole row or in the specified fields.

- `[df[ismissing.(df[!,i]), i] .= 0 for i in names(df) if Base.nonmissingtype(eltype(df[!,i])) <: Number]`: Replace `missing` with 0 values in all numeric columns, like `Float64` and `Int64`.

- `[df[ismissing.(df[!,i]), i] .= "" for i in names(df) if Base.nonmissingtype(eltype(df[!,i])) <: String]`: Replace `missing` with `""` values in all string columns.

- `isequal.(a,b)`: Compare without missing propagation (e.g., `isequal("US",missing)` returns false, while `"US" == missing` would neither be `true` nor `false` but `missing`).

- `nMissings = length(findall(x -> ismissing(x), df.col))`: Count `missing` values in a column.

9.1.8 The Split-Apply-Combine Strategy

The DataFrames package supports the common *Split-Apply-Combine* strategy through the by function, which takes in three arguments: (1) a DataFrame, (2) a column (or columns) to split the DataFrame on, and (3) a function or expression to apply to each subset of the DataFrame.

Using the by function, the "inner" function will be applied to each subgroup and whatever it returns (a scalar, an array, a named tuple, or a DataFrame) will be first horizontally merged with the values defining the subgroups and then vertically merged ("combined") to compose a final DataFrame returned by the by function. Returning a named tuple or a DataFrame allows the new columns to be labeled, with the first to be preferred for performance reasons.

One method of the by function takes instead the function as the first argument, and it's hence commonly applied using do blocks.

Internally, the by function uses the groupby() function, as in the code, it is defined as nothing else than:

```julia
by(d::AbstractDataFrame, cols, f::Function) =
combine(map(f, groupby(d, cols)))
by(f::Function, d::AbstractDataFrame, cols) = by(d, cols, f)
```

Let's see how to apply the by function to some common data transformation patterns.

Aggregating

In case of aggregation, you want to reduce the dimensionality of each subgroup defining one or more operations, each returning a scalar in place of the many possible values of the subgroup (e.g., a sum or a mean).

The inner function must return a *row*—such as a named tuple or a DataFrameRow, a DataFrame with a single line.

The following examples show the different ways to aggregate by region and by product. These examples use the prior example of the timber market data (the mean function requires the standard library package Statistics):

```julia
# Inner function returning an horizontal array (a row)
(new columns are given automatic names)
by(df, [:region, :product]) do subgroup
```

```
    [mean(subgroup.production) sum(subgroup.consumption)]
end

# Inner function returning a named tuple (returned columns are
named)
by(df, [:region, :product]) do subgroup
    (avgProduction=mean(subgroup.production),
    totalConsumption=sum(subgroup.consumption))
end

# Inner function returning a DataFrame with a single row per
subgroup
by(df, [:region, :product]) do subgroup
    DataFrame(avgProduction=mean(subgroup.production),
    totalConsumption=sum(subgroup.consumption))
end
```

As aggregating is a very common operation, the DataFrames package provides a specialized aggregate(df, aggregationColumn(s), operation(s)) function that directly aggregates the df by the required fields and applies the specified operations to each group in order to retrieve a single value for each group. For example, aggregate(df, [:region, :product], [sum, mean]) (using the earlier example) would return the sum and mean of the sets of values (production and consumption) corresponding to each category.

Note that the convenience of the aggregate function is paid by a loss in generality. You can't specify which operation to perform on the different values to aggregate (e.g., in the previous examples, the mean, just for the production, and the sum, just for the consumption).

Further, all categorical columns must be included in the parameter aggregationColumns; otherwise, Julia will try to apply the operations over them, in most cases, raising an error (as the operation expects numerical values while the categorical columns are typically strings) instead of just ignoring them.

The workaround is to remove the fields you don't need before performing the aggregation (in this example, the year: `select!(df,Not(:year)); aggregate(df, [:region, :product], [sum, mean]))`.

Computing the Cumulative Sum by Categories

Let's consider again the original (fictional) database of timber markets and seek to obtain the yearly cumulative production.

The difficulty of this task is that values on a given row depend on values of the previous rows of the same subgroup, which means that order matters.

You can use again the split-apply-combine strategy with the by function:

```julia
dfCum = by(df,[:region,:product]) do subdf
    return (year = subdf[:year], cumProd =
    cumsum(subdf[:production]))
end
```

Unlike the aggregation case, for each subgroup here, you need several rows, one for each original record. To the columns that define the subgroups (`region` and `product`), you add back the `year` and `cumsum` columns, obtained using the homologous function. Note that the DataFrames package is "smart" enough to expand the `year` scalar to an array of `cumProd` length.

Alternatively, you can use the split-apply-combine strategy with the @ linq macro (from `DataFramesMeta`):

The DataFramesMeta package provides the @linq macro, which supports a *Language Integrated Query* style over chained data transformation operations, in a way similar to R's `dplyr` package (https://cran.r-project.org/web/packages/dplyr/):

```julia
dfCum = @linq df |>
            groupby([:region,:product]) |>
            transform(cumValue = cumsum(:production))
```

Here, |> represents the *pipe* operator (we'll see it in more in detail in a later section), which passes the object returned from the operation on the left as the first argument to the operation on the right. Data (df) is initially passed to the groupby function for it to be slit, and the split subgroups are finally passed to the transform(df, operation) function to be modified.

9.1.9 Pivoting Data

Tabled data can be equivalently represented using two kinds of layouts, *long* and *wide*.

A *long* layout is when you specify each dimension on a separate column and then you have a single column for what you consider the "value". If you have multiple variables in the database, the "variable" becomes a dimensional column too.

For example, using the timber market example, Table 9-1 shows a long-formatted table.

Table 9-1. *A Long-Formatted Table*

Region	Product	Year	Variable	Value
US	Hardwood	2010	production	3.3
US	Hardwood	2010	production	3.3
US	Hardwood	2011	production	3.2
US	Softwood	2010	production	2.3
..etc..				

The long format is less human-readable, but it is very easy to work with and to perform data analysis.

In the *wide* format, on the other hand, values are expressed on multiple columns, where one or multiple dimensions are expressed along the horizontal axis. See Table 9-2.

Table 9-2. *A Widely Formatted Table*

Region	Product	Year	Production	Consumption
US	Hardwood	2010	3.3	4.3
US	Hardwood	2011	3.2	7.4
US	Softwood	2010	2.3	2.5
US	Softwood	2011	2.1	9.8
..etc..				

The timber example in the original format was a wide table, but you can create even more widely styled tables using multiple horizontal axes (not supported by the DataFrames package). See Table 9-3.

Table 9-3. *A Even More Widely Formatted Table (with Multiple Horizontal Axes)*

		Production		Consumption	
		Hardwood	Softwood	Hardwood	Softwood
US	2010	3.3	2.3	4.3	2.5
US	2011	3.2	2.1	7.4	9.8
EU	2010	2.7	1.5	3.2	6.5
EU	2011	2.8	1.3	4.3	3

The operation of moving axes between the horizontal and the vertical dimensions is called *pivoting*. In particular, *stacking* moves columns and expands them as new rows, moving hence from wide to long format. *Unstacking* does the opposite, taking rows and placing their values as new columns, thus creating a "wider" table.

Stacking Columns

The DataFrames package provides the stack(df,[cols]) function, where you specify the columns that you want to stack, and the melt(df,[cols]) function, where you specify instead the other columns, i.e., the categorical columns that are already in stacked form.

Finally, stack(df)—without column names—automatically stacks all floating-point columns.

Note that the headers of the stacked columns are inserted as data in a variable column (with the names of the variables being symbols, not strings) and the corresponding values in a value column.

```
longDf = stack(df,[:production,:consumption])                    JULIA
longDf1 = melt(df,[:region,:product,:year])
longDf2 = stack(df)
longDf == longDf1 == longDf2 # true
```

| Row | variable | value | region | product | year |
	Symbol	Float64	String	String	Int64
1	production	3.3	US	Hardwood	2010
2	production	3.2	US	Hardwood	2011
3	production	2.3	US	Softwood	2010
4	production	2.1	US	Softwood	2011

...etc...

Unstacking

The DataFrames package also provides unstack(longDf, [:rowField(s)], :variableCol, :valueCol), where you give the list of columns that should remain as categorical variables even in the wide format, the name of the column storing the category that will have to be expanded on the horizontal axis, and the name of the column containing the corresponding values.

You can omit the list of categorical variables with unstack(longDf, :variableCol, :valueCol) and then all the existing columns except the one defining column names and the one defining column values will be preserved as categorical columns:

```julia
wideDf = unstack(longDf,[:region, :product,
:year],:variable,:value)
wideDf1 = unstack(longDf,:variable,:value)
wideDf == wideDf1 # true
```

🛈 Multiple Axis Tables

The fact that the :variable parameter of unstack is a scalar and not a vector is the result of Julia DataFrames supporting only one horizontal axis.

While multiple horizontal axis DataFrames are a mess to analyze and process, they can be useful as the final presentation of the data in some contexts. You can emulate multiple horizontal axis DataFrames unstacking on different columns and then horizontally merging the sub-DataFrames. To obtain Table 9-3, you could use:

```julia
wideDf_p = unstack(wideDf,[:region,:year],:product,:production)
wideDf_c = unstack(wideDf,[:region,:year],:product,:consumption)
rename!(wideDf_p, Dict(:Hardwood => :prod_Hardwood, :Softwood =>
:prod_Softwood))
rename!(wideDf_c, Dict(:Hardwood => :cons_Hardwood, :Softwood =>
:cons_Softwood))
wideWideDf = join(wideDf_p,wideDf_c, on=[:region,:year])
```

156

The Pivot Function

The pivot(df, [:rowField(s)], :variableCol, :valueCol; <keyword arguments>) function from the LAJuliaUtils.jl package (https://github.com/sylvaticus/LAJuliaUtils.jl) can be used in order to pivot and optionally filter and sort a stacked (long) DataFrame into a single function using a spreadsheet-like pivot fashion.

pivot accepts, like unstack, the list of columns that should remain as categorical variables even in the wide format, the name of the column storing the category that will have to be expanded on the horizontal axis, and the name of the column containing the corresponding values.

The optional keywords are as follows:

- ops=sum: The operations to perform on the data, the default is to sum them. ops can be any supported Julia operation over a single array, for example: sum, mean, length, countnz, maximum, minimum, var, std, and prod. Multiple operations can be specified using an array, and in such cases, an additional op column is created to index them.

- filter: An optional filter, in the form of a dictionary of column_to_filter ⇒ [list of admissible values] (only the in filter is supported).

- sort: Optional row fields to sort. Using a tuple instead of just :colname, you can specify reverse ordering (e.g., (:colname, true)) or even a custom sort order (e.g., (:colname, [val1,val2,val3])). In such cases, the elements that you do not specify are placed at the end.

  You can pass multiple columns to be sorted in an array, e.g., [(:col1,true),:col2,(:col3,[val1, val2,val3])].

For example, the pivot function can be used with the timber long DataFrame in the following way:

```
pivDf  = pivot(longDf, [:region, :product],:variable,:value, JULIA
  ops = [mean, var],
  filter = Dict(:variable => [:consumption]),
  sort   = [:product, (:region, true)]
)
```

Region	Product	Op	Consumption
US	Hardwood	mean	5.85
US	Hardwood	var	4.805
EU	Hardwood	mean	3.75
EU	Hardwood	var	0.605
US	Softwood	mean	6.15
US	Softwood	var	26.645
EU	Softwood	mean	4.75
EU	Softwood	var	6.125

Aside from the keyword arguments, the main difference with unstack is that if you manually list the categorical columns and omit one or more column (e.g., you omit :year in wideDf = unstack(longDf,[:region, :product],:variable,:value)), you'll end up with duplicate entries (rows) in the unstacked DataFrame, with the last entries overwriting the oldest ones under the same category. With pivot, on the other hand, all entries are accounted for, using the specified aggregation operation (or sum if they aren't specified).

Sorting

DataFrames can be sorted using multiple columns with sort!(df, (:columnsToSort), rev = (reverseSortingFlags)) (for example, sort!(df,(:year, :region, :product), rev = (false, true, false))). The (optional) reverse order parameter (rev) must be a tuple of Boolean values of the same size as the cols parameter.

If you need a custom sorting of the rows, you can again rely on LAJuliaUtils, which provides customSort!(df, sortops), where sortops is the same as the sort option you saw in the pivot function.

9.1.10 Dataframe Export

Exporting to CSV

Exporting DataFrames to CSV format is directly supported by the CSV.jl package that we saw in Chapter 5:

```
CSV.write("file.csv", df)
```

Refer to section "Esxporting to CSV" in Chapter 5 for the exporting options.

Exporting to the OpenDocument Spreadsheet File

Likewise, exporting to the OpenDocument Spreadsheet format (supported by OpenOffice, LibreOffice, MS Excel, and others) is available through the OdsIO_ package (https://github.com/sylvaticus/OdsIO.jl):

```
ods_write("spreadsheet.ods",Dict( ("MyDestSheet",3,2) => df) )
```

Where the key of the dictionary is a three-element tuple that specifies the sheet name and the numerical coordinates of the top-left cell determine where to export the DataFrame.

Exporting to a Matrix

You can use `convert` to convert a database—or a portion of it—into a matrix.

- `convert(Matrix, df)`: Converts to an `Array{Any,2}`

- `convert(Matrix{Union{String,Int64,Float64}},df)`: Better, converts to an `Array{Union{Float64, Int64, String},2}`

- `convert(Matrix, df[:,[:production, :consumption]])`: Picks up homogeneous type columns, resulting in an `Array{Float64,2}`

Exporting to a Dict

To convert a DataFrame into a dictionary, you can use `toDict(df, [:dimCols], :valueCol)` provided by the `LAJuliaUtils` package, which converts a DataFrame into a dictionary by specifying the columns to be used as the key (in the given order) and the one to be used to store the value of the dictionary.

For example, `toDict(df,[:region,:product,:year],:production)` will result in the following:

```
Dict{Any,Any} with 8 entries:                          JULIA
  ("EU", "Softwood", 2011) => 1.3
  ("US", "Softwood", 2010) => 2.3
  ("US", "Hardwood", 2010) => 3.3
  ("US", "Hardwood", 2011) => 3.2
  ("EU", "Softwood", 2010) => 1.5
  ("US", "Softwood", 2011) => 2.1
  ("EU", "Hardwood", 2010) => 2.7
  ("EU", "Hardwood", 2011) => 2.8
```

Exporting to the hdf5 Format

> ℹ️ In order to use hdf5 with the HDF5 package (`https://github.com/JuliaIO/HDF5.jl`), some systems may require you to install system-wide hdf5 binaries. For example, in Deb-based Linux distributions, you should run `sudo apt-get install hdf5-tools`.

The HDF5 package doesn't support DataFrames directly, so you need first to export them as matrixes. A further limitation is that it doesn't accept a matrix of Any or Union{} types, so you may have to export the DataFrame in two pieces—the string and the numeric columns separately.

```
h5write("out.h5", "mygroup/myDf", convert(Array, df[:,[:cols]]))
```

You can read back the data by using `data = h5read("out.h5", "mygroup/myDf")`.

9.2 Using IndexedTables

IndexedTables (`https://github.com/JuliaComputing/IndexedTables.jl`) are alternative table-like data structures. Their API is somehow less convenient compared to those of DataFrames, but they are particularly interesting (a) for performant row selections, as they internally use indexed named tuples (instead of arrays in a DataFrame), and (b) for being part of a large JuliaDB ecosystem (`https://juliadb.org/`) of data-management packages, capable of handling very large sets of data that aren't held in the memory of a single machine.

They come in two variants—normal indexed tables (IndexedTable) and sparse ones (NDSparse).

Of the two, I discuss only the second one, as the first one is selectable by row position, and it's potentially less interesting than NDSparse, where you can use keys instead.

9.2.1 Creating an IndexedTable (NDSparse)

```
myTable = ndsparse((                                          JULIA
              region      = ["US","US","US","US","EU","EU","EU","EU"],
              product     = ["Hardwood","Hardwood","Softwood",
                            "Softwood","Hardwood","Hardwood",
                            "Softwood","Softwood"],
              year        = [2011,2010,2011,2010,2011,2010,2011,2010]
           ),(
             production = [3.3,3.2,2.3,2.1,2.7,2.8,1.5,1.3],
             consumption = [4.3,7.4,2.5,9.8,3.2,4.3,6.5,3.0]
           ))
```

Sparse indexed tables are created with the ndsparse function, which accepts either a named tuple (the reason for the two parentheses) or a pair of them. The reason is that IndexedTables explicitly divides between *key* columns and *value* ones (the method with a single named tuple assumes the last column as the value).

The division between key and value columns is evident when the indexed table is visualized:

```
julia> myTable                                               JULIA
3-d NDSparse with 8 values (2 field named tuples):
```

region	product	year	production	consumption
"EU"	"Hardwood"	2010	2.8	4.3
"EU"	"Hardwood"	2011	2.7	3.2
"EU"	"Softwood"	2010	1.3	3.0
"EU"	"Softwood"	2011	1.5	6.5
"US"	"Hardwood"	2010	3.2	7.4
"US"	"Hardwood"	2011	3.3	4.3
"US"	"Softwood"	2010	2.1	9.8
"US"	"Softwood"	2011	2.3	2.5

Further, you may notice that the table has been automatically sorted, like in a dictionary.

Indexed tables can also be obtained, using the same function, from an existing DataFrame, e.g., myTable = ndsparse(df).

(Sparse) indexed tables are accepted as arguments in some self-explaining functions, such as as show, first, colnames, pkeynames, and keytype.

🛑 IndexedTables are not suited for data where the "combined" key (the key across all the dimensions) is duplicated. The constructor would still let you create the indexed table, but then many methods would generate an error.

9.2.2 Row Filtering

As stated, the main advantage of indexed tables is that they provide very fast row selection.

```
t[keyForDim1,keyforDim2,:,keyForDim3,...]
```

This selects the rows matching the given set of keys, with a colon (:), indicating all values for that specific dimension. If the result is a single value, a named tuple is returned; another sparse indexed table is returned if not.

As claimed, the selection is very fast:

```
n = 1000                                                      JULIA
keys = shuffle!(reshape([[a b] for a in 1:n, b in 1:n], n^2))
# generate random keys for two dimensions.
`Shuffle` is defined in the `Random` standard library package
x = rand(n^2)
# generate random values
myTable = ndsparse((a=[keys[i][1] for i in 1:n^2],b=[keys[i][1]
for i in 1:n^2]),(x=x,))
```

```
myDf = DataFrame(a = [keys[i][1] for i in 1:n^2], b=[keys[i][1]
for i in 1:n^2], x=x)
@benchmark(myTable[100,:])              # median: 20.796 μs
@benchmark(myDf[myDf.a .==100,:])  # median:  1.297 ms
```

9.2.3 Editing/Adding Values

To change or add values, use the following:

- Change values: `myTable["EU","Hardwood",2011] = (2.8, 3.3)`

- Adding values: `myTable["EU","Hardwood",2012] = (2.8, 3.3)`

You don't need to remember the order of the value columns, as you can specify them by name:

```
myTable["EU","Hardwood",2011] = (consumption=3.4,production=2.9)
```

You *do* however have to remember the order of the key columns, as the following would not work (at the time of this writing):

```
myTable[product="Hardwood", region="EU", year=2011] =
(production=3.0, consumption=3.5) # Error !
```

9.3 Using the Pipe Operator

The `Pipe` operator (`https://github.com/oxinabox/Pipe.jl`) improves the Pipe operator `|>` in Julia Base.

Chaining (or "piping") allows you to string together multiple function calls in a way that is at the same time compact and readable. It avoids saving intermediate results without having to embed function calls within one another.

With the chain operator |>, the code to the right of |> operates on the result of the code to the left of it. In practice, the result of the operation on the left becomes the argument of the function call that is on the right.

Chaining is very useful for data manipulation. Let's assume that you want to use the following (silly) functions to operate, one after the other, on some data and print the final result:

```julia
add6(a) = a+6; div4(a) = a/4;
```
JULIA

You could either introduce temporary variables or embed the function calls:

```julia
# Method #1, temporary variables:
a = 2;
b = add6(a);
c = div4(b);
println(c) # Output: 2
# Method 2, chained function calls:
println(div4(add6(a)))
```
JULIA

With piping, you can instead write:

```julia
a |> add6 |> div4 |> println
```
JULIA

Basic piping is very limited, in the sense that it supports only functions with one argument and "piping" over only one single function at a time.

Conversely, the Pipe package provides the @pipe macro, which overrides the |> operator. It allows you to use functions with multiple arguments (and you can use the underscore character _ as a placeholder for the value of the left side). You can also pipe multiple functions in a single piping step, like so:

```julia
addX(a,x) = a+x; divY(a,y) = a/y
a = 2
# With temporary variables:
```
JULIA

165

```julia
b = addX(a,6)
c = divY(a,4)
d = b + c
println(c)
# With @pipe:
@pipe a |> addX(_,6) + divY(4,_) |> println # Output: 10.0`
```

You can even pipe over multiple values by embedding them in tuples:

```julia
data = (2,6,4)                                              JULIA
# With temporary variables:
b = addX(data[1],data[2]) # 8
c = divY(data[1],data[3]) # 0.5
d = b + c      # 8.5
println(c)
# With @pipe:
@pipe data |> addX(_[1],_[2]) + divY(_[1],_[3]) |> println
```

Note that, as with the basic pipe, functions that require a single argument that's provided by the piped data don't require parentheses.

9.4 Plotting

9.4.1 Installation and Backends

Plotting in Julia can be obtained using a specific plotting package (e.g., Gadfly at https://github.com/dcjones/Gadfly.jl, Winston at https://github.com/nolta/Winston.jl, or, as many prefer, using the Plots package (https://github.com/JuliaPlots/Plots.jl), which provides a wrapper with a unified API to several supported backends.

Backends are chosen by running chosenbackend()—that is, the name of the corresponding backend package, but written all in lowercase—before calling the plot function. The Plots package comes with GR as its default backend (see Figure 9-1), and GR.jl is installed alongside Plot.jl. If you require other backends, you can simply install the relative packages.

Two fairly complete backends are:

- Plotly.JS (https://github.com/sglyon/PlotlyJS.jl): A Julia wrapper to the plotly.js (https://plot.ly) visualization library

- PyPlot: (https://github.com/JuliaPy/PyPlot.jl): A wrapper to the Python matplotlib (http://matplotlib.org/api/pyplot_api.html)

For example:

```
using Pkg                                          JULIA
Pkg.add("Plots")
using Plots
plot(cos,-4pi,4pi, label="Cosine function (GR)") # Plot using
the default GR backend
Pkg.add("PyPlot")
Pkg.add("PlotlyJS")
Pkg.add("ORCA") # Required by PlotlyJS
pyplot()
plot(cos,-4pi,4pi, label="Cosine function (PyPlot)")
plotlyjs()
plot(cos,-4pi,4pi, label="Cosine function (PlotlyJS)")
```

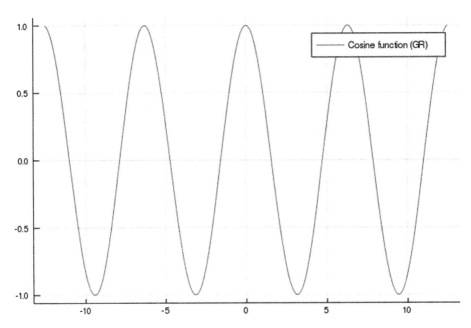

Figure 9-1. *Cosine function with the default backend GR*

While GR is the default backend, i.e., the one that will be used on every new session unless a call to other backend is made, you can set your own default setting ENV["PLOTS_DEFAULT_BACKEND"] = "myPreferredBackend" in the ~/.julia/config/startup.jl file.

🛈 Be sure not to mix different plotting packages (e.g., Plots and one of its backends used directly). This will likely create trouble. If you have imported a plot package and want to use a different one, you should always restart the Julia kernel (This is not necessary, and it is one of the advantages, when switching between different backends of the Plots package.).

You can check which backend is currently being used by Plots with backend().

There are many different backends. While basic plots can be obtained from virtually any backend, each has its own edge on some characteristic—whether it's speed, aesthetics, or interactivity, for example. Some advanced attributes may be supported only by a subset of backends.

It's important to choose the right backend for your specific case. The following pages may be of interest in this regard:

- Which backend should you choose? (http://docs.juliaplots.org/latest/backends/)

- Charts and attributes supported by the various backends (http://docs.juliaplots.org/latest/supported/)

9.4.2 The Plot Function

Plots will be rendered in the Plot panel if you're using Juno, or in a separate window if you're using the REPL (unless you use the UnicodePlots backend, which adopts characters to render the plot in a text terminal).

A relatively long first-time-to-plot is unfortunately normal.

You already saw an example of a method of the plot function, specifically plot(func,lBound,uBound;keywArgs). An other way is to plot some data using plot(xData,yData;keywArgs), where xData is an Array{T,1} of numbers or strings for the coordinates on the horizontal axis and yData is either another Array{T,1} (i.e., a column vector) or an Array{T,2} (i.e., a matrix). In the former case, a single series will be plotted; in the latter case, each matrix column would be treated as a separate data series (see Figure 9-2):

```julia
x = ["a","b","c","d","e",]
y = rand(5,3)
plot(x,y)
```

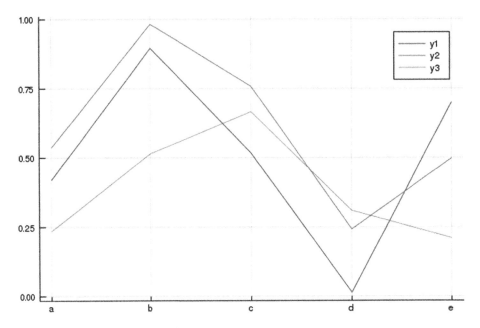

Figure 9-2. *Plotting multiple series*

While plot objects can be saved as any other object, with myPlot = plot(...), the peculiarity is that the last call to plot defines the "current" plot, which is the default to be visualized or saved. Using instead the plot! function, you add elements to an existing plot, precisely the "current" one if not otherwise specified (i.e., by indicating the plot object you are operating as the first argument).

An alternative way to plot multiple series is hence to first plot a single series, and then add one series at a time with plot!. This snippet provides a plot exactly equal to the previous one:

```
plot(x,y[:,1])  # create a new "current plot"          JULIA
plot!(x,y[:,2]) # add a serie to the current plot
plot!(x,y[:,3]) # add a serie to the current plot
```

The default type of chart drawn is a :line chart. Different types of charts can be specified with the seriestype keyword argument of the plot function, as follows:

```julia
plot(x,y[:,1]; seriestype=:scatter)
```

Common type of charts are :scatter, :line, :histogram, etc. You can obtain the full list of types supported by the current version with plotattr("seriestype").

ℹ For many types of series, plot defines alias functions like scatter(args) that correspond to plot(args, seriestype=:scatter).

Using first plot and then plot!, you can mix several types of charts together (see Figure 9-3):

```julia
plot(x,y[:,1], seriestype=:bar)
plot!(x,y[:,2], seriestype=:line)
plot!(x,y[:,3], seriestype=:scatter)
```

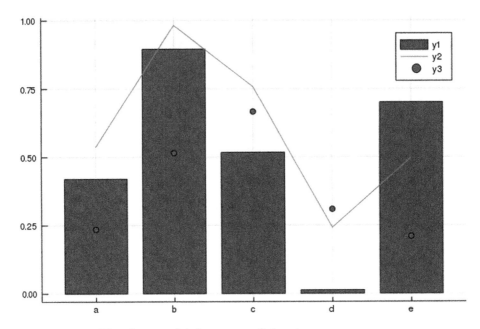

Figure 9-3. *Plotting multiple types of charts*

seriestype and label are just two of the many keyword arguments supported by plot. They are too numerous to be described in detail in this book. Refer to the Plots.jl documentation or use the inline help guide ?plot and plotattr() if you want to learn more.

9.4.3 Plotting from DataFrames

Instead of plotting data from arrays, you can plot data from DataFrames directly, using the @df macro provided by the StatsPlots package (https://github.com/JuliaPlots/StatsPlots.jl):

```
@df df plot(:year, [:production :consumption],
colour = [:red :blue])                                    JULIA
```

Using the @df macro provides three advantages:

- You don't need to repeat the df name on every series

- Series are automatically labeled with their column headers

- You can group data in the plot call itself

The following is a more elaborate example with grouped series (see Figure 9-4):

```julia
using DataFrames, StatsPlots                                    JULIA
# Let's use a modified version of our example data with more
years and just one region:
df = DataFrame(
  product        = ["Softwood","Softwood","Softwood","Softwood",
                    "Hardwood","Hardwood","Hardwood","Hardwood"],
  year        = [2010,2011,2012,2013,2010,2011,2012,2013],
  production  = [120,150,170,160,100,130,165,158],
  consumption = [70,90,100,95,  80,95,110,120]
)
pyplot()
mycolours = [:green :orange] # note it's a row vector and the
colors of the series will be alphabetically
ordered whatever order we give it here
@df df plot(:year, :production, group=:product, linestyle =
:solid, linewidth=3, label=reshape(("Production of "
.* sort(unique(:product))) ,(1,:)), color=mycolours)
@df df plot!(:year, :consumption, group=:product, linestyle =
:dot, linewidth=3, label =reshape(("Consumption of
" .* sort(unique(:product))) ,(1,:)), color=mycolours)
```

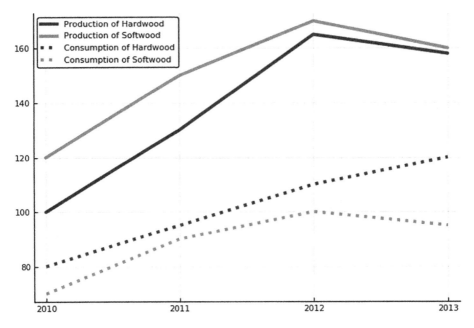

Figure 9-4. *Plotting multiple series from a DataFrame*

The key point of the previous snippet is to use `group=:product` in order to split the original series. You then use the `color` attribute to graphically separate the visualization of these new series. To differentiate between `production` and `consumption`, you use the `linestyle` attribute instead.

The legend is a bit more cumbersome to compose, as you have to consider two dimensions at the same time. In this example, you hardcode the first dimension (the market variable you are considering, either `production` or `consumption`), while you concatenate the legend item with the sorted results from `unique(:product)` for the other dimension. Finally, you reshape it as a row vector, as this is the format expected by the `label` attribute.

9.4.4 Saving

To save the currently displayed figure, you call savefig(fileName), where the extension given to fileName determines the format in which to save the figure. Some supported formats are .png (standard raster image), .pdf, and .svg (both vector formats):

```
savefig("timber_markets.svg")
savefig("timber_market.pdf")
savefig("timber_market.png")
```

JULIA

CHAPTER 10

Mathematical Libraries

This chapter includes a discussion of the following third-party packages:

Package Name	URL	Version
JuMP.jl	https://github.com/JuliaOpt/JuMP.jl	v0.19.2
GLPK.jl	https://github.com/JuliaOpt/GLPK.jl	v0.10.0
Ipopt.jl	https://github.com/JuliaOpt/Ipopt.jl	v0.5.4
SymPy.jl	https://github.com/JuliaPy/SymPy.jl	v1.0.5
LsqFit.jl	https://github.com/JuliaNLSolvers/LsqFit.jl	v0.8.1

While Julia is a complete, general-purpose programming language, it is undeniable that the focus of its community is currently concentrated on mathematical and computational questions. You already saw the large array of libraries for processing numerical data and how implications on computational performances have shaped decisions from the very beginning of Julia's development.

This chapter lists some of the most mature libraries for mathematical analyses. Some of them, like JuMP for numerical optimization, are often themselves the primary reason to start using Julia.

© Antonello Lobianco 2019
A. Lobianco, *Julia Quick Syntax Reference*, https://doi.org/10.1007/978-1-4842-5190-4_10

The list is partial and focused on relatively general tasks. A good starting point to look for specific packages is `https://pkg.julialang.org/docs/` or `https://juliaobserver.com/categories/Mathematics`.

10.1 JuMP, an Optimization Framework

JuMP (`https://github.com/JuliaOpt/JuMP.jl`) is an "Algebraic Modeling Language" (AML) for mathematical optimization problems, similar to GAMS (`https://www.gams.com/`), AMPL (`https://ampl.com/`) or Pyomo (`http://www.pyomo.org/`).

Optimization problems are problems that find the "optimal" quantities of decision variables in the context of an objective function and the constraints between the variables. Constraints arise everywhere in the real world, be it with business decisions (e.g., "How much should we produce of X?"), engineering settings (e.g., "At which angle should we design a flap for maximum efficiency?"), or personal choices (e.g., "A small flat in the city or a large house far in the countryside?").

These kinds of problems can be mathematically formulated in terms of a function to maximize (or minimize) subject to other functions acting as constraints.

Software "tools" have been created to solve these problems. They typically require you to formulate the optimization problem in a matrix format and, for non-linear problems, to provide some additional information about the problem in terms of first and second derivatives to help find the numerical optimal solution. Solver-based approaches in the MATLAB optimization toolbox or R's `optim` function are two well known examples.

AMLs take a different approach in the sense that they decouple the formulation of the problem to the specific algorithm used to solve it. The problem can now be coded algebraically, using the same mathematical notation that is used to describe it.

Critically, variables can be indexed across any dimension using the concept of "sets," thus allowing you to code equations like `Demand[year,time,region,product]` = `Supply[year,time,region,product]`. Tools that automatically provide the derivatives are included.

Once they are formulated, the problems can be solved by any "solver engine" supported by the AML software and suited for that specific problem class.

Commercial AMLs do have a drawback, however—they each use their own specific language. Being language-specific, they lag behind "real," general purpose, programming languages in many characteristics (data handling, availability of developer tools, language constructs, etc.). Interfacing them with other parts, not linked to optimization, in a larger model could be difficult.

JuMP is instead part of the "third generation" of optimization tools (the first one being the solver engines themselves and the second one being the commercial, language-specific AMLs) that are AML embedded as a software library in a general purpose programming language.

JuMP offers the same simplicity of modeling and solver-independency of language-specific AMLs while offering the benefits of a much more practical modeling environment. If it's combined with free solver engines (e.g., GLPK for Mixed Integer Problems and IPOPT for Non Linear problems), it is a complete open-source solution.

In this section, you learn how to use JuMP by implementing two problems, one with only linear functions and one where the objective is non-linear. The general settings of the problems will be the same: you choose the solver more suited to the specific problem, you load the Julia package for the solver-specific interface, you create the problem "object," and you add to it its characteristic elements, like variables, objective functions, and constraints. At this point, you can "solve" the problem and retrieve its solutions.

The examples serve also to highlight the power of Julia macros. Using them, JuMP implements its own language, while still remaining Julia.

10.1.1 The Transport Problem: A Linear Problem

The Problem

This is the transposition in JuMP of the basic transport model used in the GAMS tutorial.

Given a single product, p plants and m markets, the problem consists of defining the best routes between p and m to minimize transport costs while observing the capacity limits (a_p) of each plant and satisfying each market's demand b_m:

$$min \sum_p \sum_m c_{p,m} * x_{p,m}$$

Subject to:

$$\sum_m x_{p,m} \leq a_p$$

$$\sum_p x_{p,m} \geq b_m$$

where $c_{p,m}$ are the unitary transport costs between p and m and $x_{p,m}$ are the transported quantities (to be chosen). The original formulation is from "Dantzig G. B. (1963), *Linear Programming and Extensions, Chapter 3.3*, Princeton University Press" while its GAMS implementation can be found in "Rosenthal, R. E. (1988) *GAMS: A User's Guide, Chapter 2: A GAMS Tutorial*, The Scientific Press," or at https://www.gams.com/mccarl/trnsport.gms.

To facilitate comparison, I leave the equivalent GAMS code of each operation in the comments.

Importing the Libraries

You first need to import the JuMP package. You then need the package implementing the specific interface with the desired solver engine. In this case, we will deal with a linear problem, so you will use the *GNU Linear Programming Kit* (GLPK) library and install the GLPK.jl Julia package.

ⓘ Both the GLPK.jl and the Ipopt.jl packages will automatically install the necessary binaries for their own use (unless you direct them to use locally installed binaries, setting ENV["JULIA_GLPK_LIBRARY_PATH"], ENV["JULIA_IPOPT_LIBRARY_PATH"], and ENV["JULIA_IPOPT_EXECUTABLE_PATH"] before installing or building the packages).

The CSV package is needed to load the data.

```
using CSV, JuMP, GLPK                                            JULIA
```

Defining the Sets

JuMP doesn't have an independent syntax for sets, but it uses the native containers available in the core Julia language. Variables, parameters, and constraints can be indexed using these containers. While you can work with position-based lists, it is better to use dictionaries instead (trading out a bit of efficiency). So the "sets" in these examples are represented as lists, but then everything else (variables, constraints, and parameters) is a dictionary with the elements of the list as the keys.

```
# Define sets #                                                  JULIA
#   Sets
#       i   canning plants   / seattle, san-diego /
#       j   markets          / new-york, chicago, topeka / ;
```

```
plants  = ["seattle","san-diego"]              # canning plants
markets = ["new-york","chicago","topeka"]   # markets
```

Defining the Parameters

The capacity of plants and the demand of markets are directly defined as dictionaries, while the distance is first read as a DataFrame from a whitespace separated table. Then it is converted into a "(plant, market) ⇒ value" dictionary.

```
# Define parameters #                                       JULIA
#    Parameters
#        a(i) capacity of plant i in cases
#        /    seattle     350
#             san-diego   600  /
a = Dict(                 # capacity of plant i in cases
  "seattle"   => 350,
  "san-diego" => 600,
)

#        b(j) demand at market j in cases
#        /  new-york     325
#           chicago      300
#           topeka       275  / ;
b = Dict(                 # demand at market j in cases
  "new-york" => 325,
  "chicago"  => 300,
  "topeka"   => 275,
)

# Table d(i,j) distance in thousands of miles
#                    new-york chicago topeka
#        seattle        2.5     1.7     1.8
#        san-diego      2.5     1.8     1.4 ;
```

```
d_table = CSV.read(IOBuffer("""
plants      new-york  chicago  topeka
seattle     2.5       1.7      1.8
san-diego   2.5       1.8      1.4
"""),delim=" ", ignorerepeated=true, copycols=true)

# Here we are converting the table in a "(plant, market) =>
distance" dictionary
# r[:plants]: the first key, using the cell at the given row
and `plants` field
# m:          the second key
# r[Symbol(m)]: the value, using the cell at the given row and
the `m` field
d = Dict( (r[:plants],m) => r[Symbol(m)] for r in eachrow(d_
table), m in markets)

# Scalar f freight in dollars per case per thousand miles /90/ ;
f = 90 # freight in dollars per case per thousand miles

# Parameter c(i,j) transport cost in thousands of dollars per
case ;
#              c(i,j) = f * d(i,j) / 1000 ;
# We first declare an empty dictionary and then we fill it with
the values
c = Dict() # transport cost in thousands of dollars per case ;
[ c[p,m] = f * d[p,m] / 1000 for p in plants, m in markets]
```

Declaring the Model

Here, you declare a JuMP optimization model and give it a name. This
name will be passed as the first argument to all the subsequent operations,
like creation of variables, constraints, and objective functions (you can, if
you want, work with several models at the same time).

You can specify the solver engine at this point (as is done here) or, equivalently, at a later time when you solve the problem. You can also specify solver engine specific options.

```julia
# Model declaration (transport model)                    JULIA
trmodel = Model(with_optimizer(GLPK.Optimizer,msg_lev=GLPK.
MSG_ON)) # we choose GLPK with a verbose output
```

Declaring the Model Variables

Variables are declared and defined in the @variable macro and can have multiple dimensions—that is, they can be indexed under several indexes. Bounds are given at the same time as their declaration. Contrary to GAMS, you don't need to define the variable that is on the left side of the objective function.

```julia
## Define variables ##                                   JULIA
#  Variables
#       x(i,j)  shipment quantities in cases
#       z       total transportation costs in thousands of
#               dollars ;
#  Positive Variable x ;
@variables trmodel begin
    x[p in plants, m in markets] >= 0 # shipment quantities in
    cases
end
```

Declaring the Model Constraints

As in GAMS, each constraint can actually be a "family" of constraints, as it can be indexed over the defined sets:

```julia
## Define contraints ##                                   JULIA
# supply(i)    observe supply limit at plant i
```

```
# supply(i) .. sum (j, x(i,j)) =l= a(i)
# demand(j)   satisfy demand at market j ;
# demand(j) .. sum(i, x(i,j)) =g= b(j);
@constraints trmodel begin
    supply[p in plants],   # observe supply limit at plant p
        sum(x[p,m] for m in markets)  <= a[p]
    demand[m in markets],  # satisfy demand at market m
        sum(x[p,m] for p in plants)   >= b[m]
end
```

Declaring the Model Objective

Contrary to constraints and variables, the objective is a unique, single function. Note that it is at this point that you specify the direction of the optimization.

```
# Objective                                        JULIA
@objective trmodel Min begin
    sum(c[p,m]*x[p,m] for p in plants, m in markets)
end
```

Human-Readable Visualization of the Model (Optional)

If you want, you can get the optimization model printed in a human-readable fashion, so you can check that it's all correct.

```
print(trmodel) # The model in mathematical terms is printed JULIA
```

Resolution of the Model

It is at this point that the solver is called and the model is passed to the solver engine for its solution. Note the exclamation mark, which indicates that the trmodel object will be modified by the call.

> 💡 You can solve a model with several solver engines by making a deep copy of the model for each solver engine you want to try and indicating `optimize!` `(myCopiedModel(with_optimizer(...)))` here (Instead of in the `Model()` constructor, as you did in this example.).

```julia
optimize!(trmodel)
```

Visualization of the Results

You can check the returned status of the solver with `termination_status` `(modelObject)` (hoping for a `MOI.OPTIMAL`) and retrieve the values of the objective function and the variables with `objective_value(modelObject)` and `value(variableName)`, respectively. Duals can be obtained with a call to `dual(constraintName)`.

You can then print/visualize them as you prefer, for example:

```julia
status = termination_status(trmodel)

if (status == MOI.OPTIMAL || status == MOI.LOCALLY_SOLVED ||
status == MOI.TIME_LIMIT) && has_values(trmodel)
    if (status == MOI.OPTIMAL)
        println("** Problem solved correctly **")
    else
        println("** Problem returned a (possibly suboptimal)
        solution **")
    end
    println("- Objective value (total costs): ", objective_
    value(trmodel))
    println("- Optimal routes:")
optRoutes = value.(x)
```

```
    [println("$p --> $m: $(optRoutes[p,m])") for m in markets,
    p in plants]
    println("- Shadow prices of supply:")
    [println("$p = $(dual(supply[p]))") for p in plants]
    println("- Shadow prices of demand:")
    [println("$m = $(dual(demand[m]))") for m in markets]
else
    println("The model was not solved correctly.")
    println(status)
end
```

This should print as follows:

```
# ** Problem solved correctly **                            JULIA
# - Objective value (total costs): 153.675
# - Optimal routes:
# seattle --> new-york: 50.0
# seattle --> chicago: 300.0
# seattle --> topeka: 0.0
# san-diego --> new-york: 275.0
# san-diego --> chicago: 0.0
# san-diego --> topeka: 275.0
# - Shadow prices of supply:
# seattle = 0.0
# san-diego = 0.0
# - Shadow prices of demand:
# new-york = 0.225
# chicago = 0.153
# topeka = 0.12599999999999997
```

10.1.2 Choosing Between Pizzas and Sandwiches, a Non-Linear Problem

The only differences with non-linear problems are that you have to provide good starting points for the variables (try running the following example with 0 as a starting point!), you define the (non-linear) constraints with the specific @NLconstraints macro, and you define the (non-linear) objective with the @NLobjective macro.

Note that you don't need to provide any additional information (in particular, no derivatives are needed).

The Problem

A student has to choose how to allocate her weekly food budget of 80 dollars. She can choose only between pizzas (p) and sandwiches (s). Pizzas cost $10, while sandwiches cost only $4.

Her "utility function" (how happy she is with any given combination of pizzas and sandwiches) is $100p - 2p^2 + 70s - 2s^2 - 3ps$. She likes both, with a preference for pizza.

You use a simple problem of maximization of a non-linear function of two variables subject to a single constraint. To make the example simple, you don't use sets here (i.e., both variables and constraints are just scalar).

$$\max_{p,s} 100p - 2p^2 + 70s - 2s^2 - 3ps$$

Subject to:

$$10p + 4s \leq 80$$

Where the first equation is the utility function and the second one is what the economists call the "budget constraint," i.e., we can't spend more money than we have :-).

Importing the Libraries and Declaring the Model

As you deal with a non-linear model, you need to select a solver engine that's capable of handling this type of problem. IPOPT does the job:

```julia
using JuMP, Ipopt
m = Model(with_optimizer(Ipopt.Optimizer, print_level=0))
```

Declaring the Model Variables, Constraints, and Objectives

Since these are physical quantities, on top of the explicit constraint, you need to add a lower bound of 0 to your variables (lower and upper bounds can be specified in one expression, as lB <= var <= uB).

```julia
m = Model(with_optimizer(Ipopt.Optimizer))

@variable(m, 0 <= p, start=1, base_name="Quantities of pizzas")
@variable(m, 0 <= s, start=1, base_name="Quantities of sandwiches")

@constraint(m, budget,    10p + 4s <= 80)

@NLobjective(m, Max, 100*p - 2*p^2 + 70*s - 2*s^2 - 3*p*s)
```

You can interpret the problem as the optimal "weekly average," so it is okay if the optimal result is half a pizza. Otherwise, you should limit the variables to be integers by adding the option integer=true to their declaration. But then the class of the problem would change to Mixed Integer NonLinear Programming (MINLP) and you would have to change the solver engine as well (bonmin is good for solving MINLP problems).

Resolution of the Model and Visualization of the Results

Given that the problem is non-linear, the solution is reported as a "local optima". It is up to you then to decide if the nature of the problem guarantees that the local optima is also a global one (as in this case).

```julia
optimize!(m)

status = termination_status(m)

if (status == MOI.OPTIMAL || status == MOI.LOCALLY_SOLVED ||
status == MOI.TIME_LIMIT) && has_values(m)
    if (status == MOI.OPTIMAL)
        println("** Problem solved correctly **")
    else
        println("** Problem returned a (possibly suboptimal)
        solution **")
    end
    println("- Objective value : ", objective_value(m))
    println("- Optimal solutions:")
    println("pizzas: $(value.(p))")
    println("sandwiches: $(value.(s))")
    println("- Dual (budget): $(dual.(budget))")
else
    println("The model was not solved correctly.")
    println(status)
end

# Expected output:
# ** Problem returned a (possibly suboptimal) solution **
# - Objective value : 750.8928616403513
# - Optimal solutions:
```

```
# pizzas: 4.642857242997538
# sandwiches: 8.39285709239478
# - Dual (budget): -5.624999975248088
```

10.2 SymPy, a CAS System

Another useful mathematical package is SymPy.jl (see https://github.com/JuliaPy/SymPy.jl), a wrapper to the Python SymPy (see https://www.sympy.org/en/index.html) library for symbolic computation, that is, the *analytic* resolution of derivatives, integrals, equations (or systems of equations), and so on.

SymPy is a very large library, a *computer algebra system,* with features spanning from basic symbolic arithmetic to quantum physics. You see here the general structure of the library, solving the same problem of the previous section, but this time analytically. Refer to the SymPy documentation (https://docs.sympy.org/latest/index.html) for the complete API (much of it has a direct SymPy.jl counterpart).

10.2.1 Loading the Library and Declaring Symbols

The variables can be declared with the @var x y z syntax. You can also declare the domain of the variables, like positive, real, integer, odd, etc. (The complete list is available at https://docs.sympy.org/dev/modules/core.html#module-sympy.core.assumptions). You can solve the same example of choosing pizzas (qp) and sandwiches (qs), but keeping the price of pizza (pp) and sandwiches (ps) symbolic, just to show that the solution retrieved is analytic in nature. Only later will you obtain a specific numerical value. I simplify the example by imposing that the entire budget must be used. This allows you to use the Lagrangian multipliers method to find the analytic solution (If this is new to you, don't worry. Just follow how SymPy is used to manipulate symbolic equations and find the numerical solution later, when symbols are replaced with actual numerical values in the analytical solution.).

```
using SymPy                                                    JULIA
@vars qₚ qₛ pₚ pₛ positive=true
@vars λ
```

10.2.2 Creating and Manipulating Expressions

Once you declare the symbols, you can use them to create algebraic expressions (functions). Note that qp, qS, etc. are now both Julia variables and SymPy.jl "symbols," a reference to the underlying SymPy structure. The same is true with the expressions you create (not to be confused with Julia language expressions):

```
julia> typeof(:(qₚ+qₛ)) # A Julia language Expression   JULIA
Expr
julia> typeof(:(qₚ))     # A Julia Symbol
Symbol
julia> typeof(qₚ+qₛ)     # A SymPy expression, i.e. a SymPy object
Sym
```

You can create the so-called "Lagrangian" by adding to the objective solution each constraint multiplied by the relative "Lagrangian multiplier".

You can then retrieve the partial derivatives of this Lagrangian function using diff(function,variable):

```
utility = 100*qₚ - 2*qₚ^2 + 70*qₛ - 2*qₛ^2 - 3*qₚ*qₛ   JULIA
budget  = pₚ* qₚ + pₛ*qₛ
lagr    = utility + λ*(80 - budget)
dlqₚ    = diff(lagr,qₚ)
dlqₛ    = diff(lagr,qₛ)
dlλ     = diff(lagr,λ)
```

As claimed, each derivative is expressed in symbolic terms:

```julia
julia> dlqₚ                                              JULIA
-pₚ·λ - 4·qₚ - 3·qₛ + 100
```

10.2.3 Solving a System of Equations

The first order conditions tell you that the solutions, expressed in terms
of qp, qs, and λ, will be found by setting the relative derivatives of the
Lagrangian equal to zero. That is, you solve a system of equations of three
variables in three unknowns.

solve([equations],[variables]) does exactly that:

```julia
julia> sol = solve([Eq(dlqₚ,0), Eq(dlqₛ,0), Eq(dlλ,0)],
[qₚ, qₛ, λ]) # SymPy.solve if a function named "solve"
has been already defined in other packages                JULIA
Dict{Any,Any} with 3 entries:
  λ => 5*(19*pₚ - 2*pₛ - 56)/(2*pₚ^2 - 3*pₚ*pₛ + 2*pₛ^2)
  qₚ => 5*(-7*pₚ*pₛ + 32*pₚ + 10*pₛ^2 - 24*pₛ)/(2*pₚ^2 - 3*pₚ*pₛ + 2*pₛ^2)
  qₛ => 5*(7*pₚ^2 - 10*pₚ*pₛ - 24*pₚ + 32*pₛ)/(2*pₚ^2 - 3*pₚ*pₛ + 2*pₛ^2)
```

💡 The solve function also accepts expressions directly, setting
them in that case equal to zero. Hence, this would have also worked:
solve([dlqp, dlqs, dlλ], [qp, qs, λ]).

10.2.4 Retrieving Numerical Values

You can use myExpression.evalf(subs=Dict(symbol=>value)) to "inject"
into the expression the given numerical values of its symbol and retrieve
the corresponding numerical value of the expression.

You retrieve the numerical values of the solutions by specifying the prices of pizzas as $10 and sandwiches as $4:

```julia
qₚ_num = sol[qₚ].evalf(subs=Dict(pₚ=>10,pₛ=>4)) # 4.64285714285714
```

```julia
qₛ_num = sol[qₛ].evalf(subs=Dict(pₚ=>10,pₛ=>4)) # 8.39285714285714
λ_num = sol[λ].evalf(subs=Dict(pₚ=>10,pₛ=>4)) # 5.625
```

> ⓘ Even if they're visualized as numbers, q_p_num, q_s_num, and λ_num remain SymPy objects. To convert them to Julia numbers, you can use N(SymPySymbol). For example, N(q_p_num).

Finally, to retrieve the numerical value of the objective function, let's substitute the symbols in the utility function with these numerical values:

```julia
z = utility.evalf(subs=Dict(qₚ=>qₚ_num, qₛ=>qₛ_num)) #750.892857142857
```

You have retrieved the same "optimal" values you found numerically with JuMP (as expected).

10.3 LsqFit, a Data Fit Library

As a third example, I selected a data fit model. Statistics (and the related field of machine learning) is possibly the area with the most active development in the Julia community. The LsqFit package offers lots of flexibility while remaining user-friendly.

You will use it to estimate the logistic growth curve (Verhulst model) of a forest stand of beech (a broadleaved species) given a sample of timber volume data from a stand in northwest France[1].

---

[1]Data is from ENGREF, (1984) *Tables de production pour les forets francaises*, 2e édition, p. 80. Age is given in years, volumes in metric cubes per hectare.

10.3.1 Loading the Libraries and Defining the Model

You start by specifying the functional form of the model you want to fit and where the parameters to be estimated stand. The @. macro is a useful tool to avoid specifying the dot operator (for broadcasting) on each operation.

The model you want to estimate is the well-known logistic function:

$$vol = \frac{par_1}{1+e^{-par_2*(age-par_3)}}$$

```
using LsqFit,CSV,Plots                                          JULIA

# **** Fit volumes to logistic model ***
@. model(age, pars) = pars[1] / (1+exp(-pars[2] * (age - pars[3])))
obsVols = CSV.read(IOBuffer("""
age   vol
20    56
35    140
60    301
90    420
145   502
"""), delim=" ", ignorerepeated=true, copycols=true)
```

10.3.2 Parameters

Since the algorithm to find the fit is non-linear, you must provide some "reasonable" starting points. In the case of the logistic functions, the first parameter is the maximum level reached by the function (the "carrying capacity"), the second is the growth rate, and the third is the midpoint on the x scale. The following could be reasonable starting points:

```
par0 = [600, 0.02, 60]                                          JULIA
```

10.3.3 Fitting the Model

Once the model has been defined, you can fit it with your data and your initial guess of the parameters (you could have also specified lower and upper bounds of the parameters and then passed them as `lower=lb`, `upper=ub`). This is the stage where the least-squares fitting happens.

```
fit = curve_fit(model, obsVols.age, obsVols.vol, par0)       JULIA
```

10.3.4 Retrieving the Parameters and Comparing them with the Observations

Fitted parameters are now available in the `fit.param` array. At this point, you can simply use them to compute the fitted volumes across every year and compare them with the observed values (see Figure 10-1):

```
fit.param # [497.07, 0.05, 53.5]                             JULIA
fitX = 0:maximum(obsVols.age)*1.2
fitY = [fit.param[1] / (1+exp(-fit.param[2] * (x - fit.
param[3]))) for x in fitX]
plot(fitX,fitY, seriestype=:line, label="Fitted values")
plot!(obsVols.age, obsVols.vol, seriestype=:scatter, label="Obs
values")
plot!(obsVols.age, fit.resid, seriestype=:bar,
label="Residuals")
```

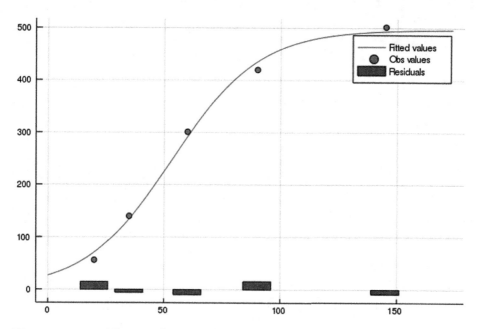

Figure 10-1. *Observed versus fitted data*

CHAPTER 11

Utilities

We conclude this book with a discussion of the following third-party packages, as laid out in the following table:

Package Name	URL	Version
Weave.jl	https://github.com/mpastell/weave.jl	v0.9.1
ZipFile.jl	https://github.com/fhs/ZipFile.jl	v0.8.3
Interact.jl	https://github.com/JuliaGizmos/Interact.jl	v0.10.2
Mux.jl	https://github.com/JuliaWeb/Mux.jl	v0.7.0

I also mention a few additional Julia packages that are of more general use.

11.1 Weave for Dynamic Documents

Weave (https://github.com/mpastell/weave.jl) allows developers to produce so-called "dynamic documents," where the code producing the "results" is embedded directly into the document. The overall structure of the document is a *markdown* document, while the Julia code is inserted as *code cells* within the main document.

In many situations, this is very useful, as it enables the logic going from the assumptions to the results and the presentation and discussion of these results to be presented in the same document.

© Antonello Lobianco 2019
A. Lobianco, *Julia Quick Syntax Reference*, https://doi.org/10.1007/978-1-4842-5190-4_11

Weave, mixing the code and the (markdown-based) documentation, is similar to a Jupyter Notebook. Its main advantage is that the code cells are not defined *a priori*, but you can still run the part of the code that you need in Juno, independent of the way you organized the cells in the document. This is thanks to the Atom package called *language-weave* (https://atom. io/packages/language-weave).

To see an example, consider the following document as being a file with the .jmd extension (e.g., testWeave.jmd):

```                                                                JULIA
---
title :         Test of a document with embedded Julia code and
                citations
date :          5th September 2019
bibliography: biblio.bib
---


# Section 1 (leave two rows from the document headers above)

Weave.jl, announced in @Pastell:2017, is a scientific report
generator/literate programming tool for Julia
developed by Matti Pastell, resembling Knitr for R [see @
Xie:2015].

## Subsection 1.1

This should print a plot. Note that, with `echo=false`, we are
not rendering the source code in the final PDF:

```{julia;echo=false}
using Plots
pyplot()
plot(sin, -2pi, pi, label="sine function")
```
```

Here instead we will render in the PDF both the script source code and its output:

```
```{julia;}
using DataFrames
df = DataFrame(
        colour = ["green","blue","white","green","green"],
        shape  = ["circle", "triangle", "square","square","circle"],
        border = ["dotted", "line", "line", "line", "dotted"],
        area   = [1.1, 2.3, 3.1, missing, 5.2]
    )
df
```
```

Note also that we can refer to variables defined in previous chunks (or "cells", following Jupyter terminology):

```
```{julia;}
df.colour
```
```

Subsubsection

For a much more complete example see the [Weave documentation] (http://weavejl.mpastell.com/stable/).

References

While the previous document is a Weave *Julia markdown* (.jmd) document, the following `biblio.bib` file is a standard BibTeX file containing the necessary references:

```
@article{  Pastell:2017,
  author  = {Pastell, Matti},
  title   = {Weave.jl: Scientific Reports Using Julia},
```

LATEX

```
  journal = {Journal of Open Source Software},
  vol     = {2},
  issue   = {11},
  year    = {2017},
  doi     = {10.21105/joss.00204}
}
@Book{       Xie:2015,
  title     = {Dynamic Documents with R and Knitr.},
  publisher = {Chapman and Hall/CRC},
  year      = {2015},
  author    = {Yihui Xie},
  edition   = {2nd ed},
  url       = {http://yihui.name/knitr/},
}
```

If you are in Juno, you can keep testWeave.jmd open in the main panel and "play" with the Julia code. When you are satisfied, you can compile the document by running the following commands from the Julia console in Juno:

```
julia> using Weave;                                           JULIA
julia> weave("testWeave.jmd", out_path = :pwd)
julia> weave("testWeave.jmd", out_path = :pwd, doctype =
"pandoc2pdf")
```

Conditional to the presence in the system of the necessary toolss[1], the first command would produce an HTML document, while the second one would produce the PDF shown in Figure 11-1.

---

[1]In Ubuntu Linux (but most likely also in other systems), Weave needs Pandora >= 1.20 and LaTeX (texlive-xetex) to be installed on the system. If you use Ubuntu, the version of Pandora in the official repositories may be too old. Use the deb available at https://github.com/jgm/pandoc/releases/latest instead.

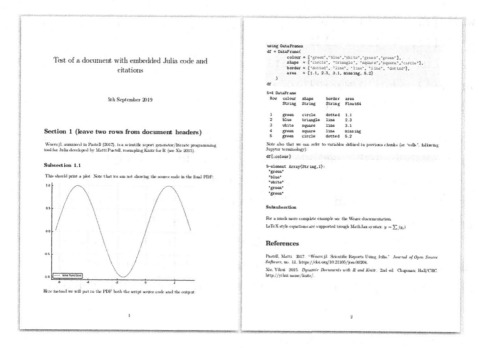

Figure 11-1. *Rendered wave PDF*

11.2 Zip Files

At one point or another, we all have to deal with zip archives.

ZipFile (https://github.com/fhs/ZipFile.jl) provides a library to manage zip archives easily.

11.2.1 Writing a Zip Archive

You first open the zip file acting as container. Then, you add the files you need and write over them. It's important to close the file, as the data is flushed on disk (at least for small files) and the zip file is finalized at that point.

```julia
zf = ZipFile.Writer("example.zip")
f1 = ZipFile.addfile(zf, "file1.txt", method=ZipFile.Deflate)
```

```julia
write(f1, "Hello world1!\n")
write(f1, "Hello world1 again!\n")
f2 = ZipFile.addfile(zf, "dir1/file2.txt", method=ZipFile.Deflate)
write(f2, "Hello world2!\n")
close(zf) # Important!
```

The package does not export Writer() or addfile(), so you need to prepend them to the package name, as shown in the example.

Also, when you add a file, you need to specify method=ZipFile. Deflate. Otherwise, the file will be stored uncompressed in the zip archive, and that is not what you want in most situations.

In the previous example, you used two different objects for the two files. Note that a bug in the library prevents you from adding all the files and then writing on them in a second step. For example, this would not work:

```julia
zf = ZipFile.Writer("example.zip")
f1 = ZipFile.addfile(zf, "file1.txt", method=ZipFile.Deflate)
f2 = ZipFile.addfile(zf, "dir1/file2.txt", method=ZipFile.Deflate)
write(f1, "Hello world1!\n") # Error !
write(f1, "Hello world1 again!\n")
write(f2, "Hello world2!\n")
close(zf) # Important!
```

11.2.2 Reading from a Zipped Archive

The process for reading a zip archive is similar. First, you open the archive in reading mode with ZipFile.Reader (again, you need to use the function name prepended with the package name) and then you can loop over its files property and read from it, either using eachline or read.

You don't need to close each individual file, but you do need to close the zip archive:

```julia
zf = ZipFile.Reader("example.zip");                         JULIA
for f in zf.files
    println("*** $(f.name) ***")
    for ln in eachline(f) # Alternative: read(f,String) to read
    the whole file
        println(ln)
    end
end
close(zf) # Important!
```

You can also retrieve other information concerning the file by looking at its name, method (Store|Deflate), uncompressedsize, and compressedsize properties (e.g., f.name).

To get information about the overall zip archive, use show(zf) instead:

```julia
julia> zf = ZipFile.Reader("example.zip");                  JULIA
julia> show(zf)
ZipFile.Reader for IOStream(<file 11_example.zip>) containing 2
files:
uncompressedsize method mtime                   name
-------------------------------------------------
              34 Deflate 2019-07-18 12-06 file1.txt
              14 Deflate 2019-07-18 12-06 dir1/file2.txt
julia> close(zf)
```

11.3 Interact and Mux: Expose Interacting Models on the Web

When you need to put a model into production, letting the user choose the parameters and visualize the result, you can employ Interact.jl (https://github.com/JuliaGizmos/Interact.jl) to make the widgets and Mux.jl (https://github.com/JuliaWeb/Mux.jl) to embed them in a web page.

The result is very similar to *Shiny for R* (https://shiny.rstudio.com/).

11.3.1 Importing the Libraries

As mentioned, you need Interact to draw the widgets and set them down in a layout, Plots to actually draw the plot (one of the widgets you will create in the example), and Mux to initiate a web server and serve the layout to the browser's requests.

```julia
using Interact,Plots,Mux
```

11.3.2 Defining the Logic of the Model

Here, we wrap the model in a function that will receive the parameters from the controls and produce the "results" to be visualized in a plot.

```julia
function myModel(p1,p2)
    xrange = collect(-50:+50)
    model = p1.*xrange .+ p2.* (xrange) .^2
    return model
end
```

11.3.3 Defining Controls and Layout

You can now create the widgets: two controls for the two parameters of the model and one plot to display the results. Under the hood, when the sliders move, the model receives the new set of parameters, new results are computed, and the plots figure is updated.

You start by creating the two sliders, providing them with a range and a default value. Of course, other types of widgets are available (e.g., a direct numeric input without a range, or selectable lists).

```julia
function createLayout()
  p1s = slider(-50:50, label = "Par#1 (linear term):", value = 1)
  p2s = slider(-5:0.1:5, label = "Par#2 (quad term):", value = 1)
```

You now use the Interact.@map macro to update the output as soon as you change the sliders. Note the ampersand (&) syntax is provided by the macro to resemble C references, in order to indicate the elements of the given expression that have to be "observed" for changes.

The output of the model is a vector of Float64. The @map macro wraps it in another Observable object, mOutput. This is chained as input to the plot function, where plt is another Observable.

Finally, the two sliders are grouped in a custom widget using the Widget constructor and put together with the plot in a common layout defined by the @layout! macro.

```julia
  mOutput = Interact.@map myModel(&p1s,&p2s)
  plt = Interact.@map plot(collect(-50:50),&mOutput,
  label="Model output")
  wdg = Widget(["p1" => p1s, "p2" => p2s], output = mOutput)
  @layout! wdg hbox(plt, vbox(:p1, :p2))
end
```

If Juno is used, you can test it with a call to createLayout(). After a slight delay, the layout should be displayed in the Juno plot pane, and you can play with it.

11.3.4 Providing Widgets to Web Users

If you are satisfied with the job, you can finally expose it to the web. With WebIO.webio_serve(page("destinationPath", req -> interactiveWidget), destinationPort) (from the Mux.jl package), you create a server that answers browser requests at the given port and path serving the widget you made.

You wrap it all in an exception-handling block in order to avoid recoverable connection errors linked to timeouts being set too small:

```julia
function serveLayout(destinationPort)
    try
      WebIO.webio_serve(page("/", req -> createLayout()),
      destinationPort)
    catch e
      if isa(e, IOError)
        # sleep and then try again
        sleep(0.1)
        serveLayout(destinationPort)
      else
        throw(e)
      end
    end
end

serveLayout(8001)
```

Finally, you can connect to the server at the given port by using a web browser and pointing it to `http://localhost:destinationPort`, as shown in Figure 11-2.

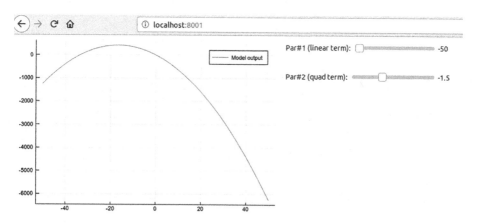

Figure 11-2. *The interactive plot rendered in a web browser*

The connection would use the HTTP protocol. HTTP over TLS (`https://foo`) can be obtained using a tunnel, e.g., by employing the Apache proxy server.

Index

A

Abstract Syntax Tree (AST), 82, 83
Abstract types, 61
    inheritance, 66
    MyOwnAbstractType, 63
    MyOwnGenericAbstract
        Type, 62
    object-oriented
        composition, 64
        inheritance, 63
        multiple dispatch, 65
Apache proxy server, 209
Arrays
    a[mask], 28
    creation, 21, 22
    from:step:to syntax, 22
    functions, 23–25
    multi-dimensional, 26, 27
    n-dimensional, 28–30
    nested, 27
    union keyword, 22

B

Benchmarking
    @benchmark, 116
    BenchmarkTools package, 116
    Fibonacci sequence, 115
    JIT, 115
    @time macro, 115
Budget constraint, 188

C

Code block structure
    flow-control constructs, 41
    global variables, 42, 43
Comma separated value
    (CSV), 70, 71, 77
Conditional statements, 44–45
createLayout(), 208
Cython, 3

D

Dataframe export
    to CSV, 159
    to dictionary, 160
    HDF5 package, 161
    to matrix, 160
    OpenDocument
        Spreadsheet, 159
DataFrames
    create/load data, 139–141
    data insights, 141, 142
    defined, 138

© Antonello Lobianco 2019
A. Lobianco, *Julia Quick Syntax Reference*, https://doi.org/10.1007/978-1-4842-5190-4

Printed in the United States
By Bookmasters